Research

Director, Fiscal and Non-Profit Studies, Jason Clemens
Director, School Performance Studies, Peter Cowley
Acting Director, Pharmaceutical Policy Research, John R. Graham
Director, Centre for Studies in Risk and Regulation, Laura Jones
Director, Social Affairs Centre, Fred McMahon
Director, Education Policy, Claudia Rebanks Hepburn
Senior Research Economist, Joel Emes

Ordering publications

To order this book, any other publications, or a catalogue of the Institute's publications, please contact the book sales coordinator via our **toll-free order line: 1.800.665.3558, ext. 580**; via telephone: 604.688.0221, ext. 580; via fax: 604.688.8539; via e-mail: sales@fraserinstitute.ca.

Media

For media information, please contact Suzanne Walters, Director of Communications: via telephone: 604.714.4582 or, from Toronto, 416.363.6575, ext. 582; via e-mail: suzannew@fraserinstitute.ca

Website

To learn more about the Institute and to read our publications on line, please visit our web site at www.fraserinstitute.ca.

Membership

For information about membership: in **Vancouver**, please contact us via mail: The Development Department, The Fraser Institute, 4th Floor, 1770 Burrard Street, Vancouver, BC, V6J 3G7; via telephone: 604.688.0221 ext. 586; via fax: 604.688.8539; via e-mail: membership@fraserinstitute.ca

In **Calgary**, please contact us via telephone: 403.216.7175; via fax: 403.234.9010; via e-mail: paulineh@fraserinstitute.ca

In **Toronto**, please contact us via telephone: 416.363.6575; via fax: 416.601.7322.

Publication

Editing and design by Kristin McCahon and Lindsey Thomas Martin
Cover design by Brian Creswick @ GoggleBox.

Risk Controversy Series
General Editor, Laura Jones

The Fraser Institute's Risk Controversy Series publishes a number of short books explaining the science behind today's most pressing public-policy issues, such as global warming, genetic engineering, use of chemicals, and drug approvals. These issues have two common characteristics: they involve complex science and they are controversial, attracting the attention of activists and media. Good policy is based on sound science and sound economics. The purpose of the Risk Controversy Series is to promote good policy by providing Canadians with information from scientists about the complex science involved in many of today's important policy debates. The books in the series are full of valuable information and will provide the interested citizen with a basic understanding of the state of the science, including the many questions that remain unanswered.

Upcoming issues of the Risk Controversy Series will investigate genetically modified food and misconceptions about the causes of cancer. Suggestions for other topics are welcome.

About the Centre for Studies in Risk and Regulation

The Fraser Institute's Centre for Studies in Risk and Regulation aims to educate Canadian citizens and policy-makers about the science and economics behind risk controversies. As incomes and living standards have increased, tolerance for the risks associated with everyday activities has decreased.

While this decreased tolerance for risk is not undesirable, it has made us susceptible to unsound science. Concern over smaller and smaller risks, both real and imagined, has led us to demand more regulation without taking account of the costs, including foregone opportunities to reduce more threatening risks. If the costs of policies intended to reduce risks are not accounted for, there is a danger that well-intentioned policies will actually reduce public well-being. To promote more rational decision-making, the Centre for Studies in Risk and Regulation will focus on sound science and consider the costs as well as the benefits of policies intended to protect Canadians.

For more information about the Centre, contact Laura Jones, Director, Centre for Studies in Risk and Regulation, The Fraser Institute, Fourth Floor, 1770 Burrard Street, Vancouver, BC, V6J 3G7; via telephone: 604.714.4547; via fax: 604.688.8539; via e-mail: lauraj@fraserinstitute.ca

Risk Controversy Series 1

Global Warming
A Guide to the Science

Willie Soon,
Sallie L. Baliunas,
Arthur B. Robinson and
Zachary W. Robinson

 The Fraser Institute
Centre for Studies in Risk and Regulation
Vancouver British Columbia Canada 2001

Printed in Canada.

National Library of Canada Cataloguing in Publication Data

Main entry under title:
Global warming

Risk controversy series; no. 1
Includes bibliographical references.
ISBN 0-88975-187-0

1. Global warming. 2. Greenhouse effect, Atmospheric.
I. Soon, Willie. II. Fraser Institute (Vancouver, B.C.) III. Series.
QC981.8.G56G55 2001   363.738'74   C2001-911355-2

Contents

About the authors

Sallie Baliunas is an astrophysicist at the Harvard-Smithsonian Center for Astrophysics in Cambridge, Massachusetts. She is also the deputy director of Mount Wilson Observatory, senior scientist at the George C. Marshall Institute and Visiting Professor at Brigham Young University. She is also co-host of Tech Central Station.com.

Art Robinson is President and Research Professor of Chemistry at the Oregon Institute of Science and Medicine (OISM). He holds a BS in chemistry from the California Institute of Technology and a PhD in chemistry from the University of California at San Diego, where he was appointed to the faculty immediately after receiving his PhD. Later, he served as President and Research Director of the Linus Pauling Institute of Science and Medicine, before founding OISM.

Zachary Robinson holds a BS in chemistry from Oregon State University. He is currently working simultaneously for a PhD in Chemistry and a DVM in Veterinary Medicine at Iowa State University.

Willie Soon is a physicist at the Harvard-Smithsonian Center for Astrophysics in Cambridge, Massachusetts. He has served as an astronomer at Mount Wilson Observatory since 1992 and is a senior scientist at the George C. Marshall Institute.

Foreword

Global Warming: A Guide to the Science is the first publication in The Centre for Studies in Risk and Regulation's Risk Controversy Series, which will explain the science behind many of today's most pressing public-policy issues. Many current public-policy issues such as global warming, genetic engineering, use of chemicals, and drug approvals have two common characteristics: they involve complex science and they are controversial, attracting the attention of activists and media. The mix of complex science, activists' hype, and short media clips can bewilder the concerned citizen.

The activists
The development and use of new technology has long attracted an "anti" movement. Recent high-profile campaigns include those against globalization, genetic engineering, cell phones, breast implants, greenhouse gases, and plastic softeners used in children's toys. To convince people that the risks from these products or technologies warrant attention, activists rely on dramatic pictures, public protests, and slogans to attract media attention and capture the public's imagination. The goal of these campaigns is not to educate people so they can make informed choices for themselves—the goal is to regulate or, preferably, to eliminate the offending product or technology. While activists' personal motivations vary, their campaigns have three common characteristics. First, there is an underlying suspicion of economic development. Many prominent environmental activists, for example, say that economic growth is the enemy of the environment and among anti-globalization crusaders, "multinational corporation" is a dirty word. Second, the benefits of the products, technologies, or lifestyles that activists attack are ignored while the risks are emphasized and often exaggerated. Many activists insist that a product or technology be proven to pose no risk at all

before it is brought to market—this is sometimes called the precautionary principle. This may sound sensible but it is, in fact, an absurd demand: nothing, including many products that we use and activities we enjoy daily, is completely safe. Even the simple act of eating an apple poses some risk—one could choke on the apple or the apple might damage a tooth. Finally, activists have a tendency to focus only on arguments that support their claims, which often means dismissing legitimate scientific debates and ignoring uncertainty: activists claim, for example, that there is a consensus among scientists that global warming is caused largely by human activity and that something must therefore be done to control greenhouse gas emissions. As this publication shows, no such consensus exists.

The media

Many of us rely exclusively on the media for information on topics of current interest as, understandably, we do not have time to conduct our own, more thorough literature reviews and investigations. For business and political news as well as for human-interest stories, newspaper, radio, and television media do a good job of keeping us informed. But, these topics are relatively straight-forward to cover as they involve familiar people, terms, and places. Stories involving complex science are harder to do. Journalists covering these stories often do not have a scientific background and, even with a scientific background, it is difficult to condense and simplify scientific issues for viewers or readers. Finally, journalists work on tight deadlines, often having less than a day to research and write a story. Tight deadlines also make it tempting to rely on activists who are eager to provide information and colourful quotations.

Relying on media for information about a complex scientific issue can also give one an unbalanced view of the question because bad news is a better story than good news. In his book, *A Moment on the Earth*, Gregg Easterbrook, a reporter who has covered environmental issues for *Newsweek, The New Republic,* and *The New York Times Magazine,* explains the asymmetry in the way the media cover environmental stories.

> In the autumn of 1992, I was struck by this headline in the *New York Times*: "Air Found Cleaner in US Cities." The accompanying story said that in the past five years air quality had improved sufficiently that nearly half the cities once violating federal smog standards no longer did so. I was also struck by how the *Times* treated the article—as a small box buried on page A24. I checked the nation's other important news organizations and learned that none had given the finding prominence. Surely any news that air quality was in decline would have received front-page attention (p. xiii).

Despite dramatic overall improvements in air quality in Canada over the past 30 years, stories about air quality in Canada also focus on the bad news. Both the *Globe and Mail* and the *National Post* emphasized reports that air quality was deteriorating. Eighty-nine percent of the *Globe and Mail*'s coverage of air quality and 81 percent of the *National Post*'s stories in 2000 focused on poor air quality (Miljan, Air Quality Improving—But You'd Never Know It from the *Globe & Post, Fraser Forum,* April 2001: 17–18).

That bad news makes a better story than good news is a more generally observable phenomenon. According to the Pew Research Center for the People and the Press, each of the top 10 stories of public interest in the United States during 1999 were about bad news. With the exception of the outcome of the American election, the birth of septuplets in Iowa, and the summer Olympics, the same is true for the top 10 stories in each year from 1996 through 1998 (Pew Research Center for the People and the Press 2000, digital document: www.people-press.org/yearendrpt.htm).

While it is tempting to blame the media for over-simplifying complicated scientific ideas and presenting only the bad news, we must remember that they are catering to the desires of their readers and viewers. Most of us rely on newspapers, radio, and television because we want simple, interesting stories. We also find bad news more interesting than good news. Who would buy a paper that had "Millions of Airplanes land safely in Canada each Year" as its headline? But, many of us are drawn to headlines that promise a story giving gory details of a plane crash.

The Risk Controversy Series

Good policy is based on sound science and sound economics. The purpose of the Risk Controversy Series is to promote good policy by providing Canadians with information from scientists about the complex science involved in many of today's important policy debates. While these reports are not as short or as easy to read as a news story, they are full of valuable information and will provide the interested citizen with a basic understanding of the state of the science, including the many questions that remain unanswered.

Upcoming issues of the Risk Controversy Series will investigate genetically modified food and misconceptions about the causes of cancer. Suggestions for other topics are welcome.

Laura Jones, Director
Environment and Regulatory Studies
Centre for Studies in Risk and Regulation

Global Warming
A Guide to the Science

Abstract

A review of the scientific literature concerning the environmental consequences of increased levels of atmospheric carbon dioxide, the most prominent greenhouse gas contributed by human activities, leads to the conclusion that increases during the twentieth century have produced no deleterious effects upon global climate or temperature. Increased carbon dioxide has, however, markedly increased the growth rates of plants as inferred from numerous laboratory and field experiments. There is no clear evidence, nor unique attribution, of the global effects of anthropogenic CO_2 on climate. Meaningful assessments of the environmental impacts of anthropogenic CO_2 are not yet possible because model estimates of global and regional changes in climate on interannual, decadal and centennial time-scales remain highly uncertain.

The Glossary
Technical terms marked with emphasis in the text (e.g. *General Circulation Model*) are explained in the Glossary (pages 42–44).

Summary

The earth's atmosphere contains greenhouse gases that absorb some of the energy that, in the absence of those gases, would escape to space. The absorbing property of those gases in the atmosphere help create the greenhouse effect, which makes the earth warmer than it would be if those gases were not present in the air. Most of the greenhouse effect arises from water vapor in the air and water in clouds, with minor contributions from, for example, carbon dioxide and methane. In contrast, the gases nitrogen and oxygen, which make up most of the earth's atmosphere, lack the property of absorbing infrared radiation that characterizes a greenhouse gas.

Since the start of the Industrial Revolution, human activities like the burning of coal or oil have significantly raised the carbon-dioxide content of the air. This increase should warm the earth and produce an enhanced greenhouse effect.

The temperature at the surface of the earth, measured over the last 150 years or so by thermometers placed on land and sea at locations scattered over the globe, has been rising. The compilation of over 70 million thermometer readings, plus signals contained in mountain glaciers, tree-growth rings, coral layers and other biological or geological indicators that are sensitive to temperature change, agree that the twentieth century saw a global warming.

Do these two facts taken together mean that the human release of carbon dioxide caused the warming observed during the twentieth-century? If so, will a future warming trend lead to a climatic catastrophe owing to the use of carbon-based energy?

Reality tells a different story. The historical surface and proxy records suggest that temperatures rose about 0.5°C in the early twentieth century—before most of the greenhouse gases were added to the air by human activities.

The warming in the early twentieth-century must have been natural. The surface temperature peaked by around 1940, then cooled until the 1970s. Since then, there has been a surface warming. Little, if any, of the twentieth-century warming can be attributed to the recent rise in the carbon dioxide content of the air.

However, the expected continued increase in the concentration of carbon dioxide in the air leads to concern for a disastrous rise in the global temperature in the future. This concern mainly stems from computer-based simulations of the climate system, forecast through the next century. The common tool for a computer simulation of the climate is the **General Circulation Model** (GCM). The climate models are an integral part not only of the science of climate change but also of the policy debate. Thus, an important question is: how good are the models in forecasting future climate change?

The climate of the earth is dynamic and includes phenomena that range in size from molecules and particle droplets in clouds to wind patterns over a hemisphere. That means climatic phenomena range in scale over roughly 16 powers of 10. Also, many climatic processes interact with each other in complex ways; many are still mysterious as, for example, the way sunlight at different wavelengths interacts with clouds.

At any moment, around five million different variables have to be followed in a computer mock-up of the climate. All their important impacts and interactions must be known, yet it is certain that they are not all known. It is not surprising, therefore, that to calculate reliably the climatic impacts of increases in the concentration of atmospheric carbon dioxide remains very difficult.

Major components of the climatic system are not satisfactorily represented in the models because there is a lack of good understanding of climate dynamics, both on theoretical and observational grounds. The models give a range of outcomes. Typically, the aggregate outcome of various GCMs is listed as a 1.5°C to 4.5°C rise in global temperature for an approximate doubling of the concentration of atmospheric CO_2 (Houghton & al. 1996). The agreement of the outcomes and the range of changes produced by the models are not to be taken literally. The results from the models do not constitute a statistical or physical mean and standard deviation. Given the substantial uncertainties associated with the modeling enterprise and its many *parameterizations*, the outcomes of the models, which are subject to large systematic errors, cannot be averaged and represented as a consensus result.

Further, there are independent, semi-empirical approaches that give results lying outside the range of temperature change normally output by the models after parameterization. For example, analysis of the climatic response to perturbations by volcanic eruptions suggests a climatic sensitivity of 0.3°C to 0.5°C for a doubling of atmospheric CO_2 (Lindzen 1997). In addition, consideration of a variety of biological and other negative feedbacks in the climatic system yields a climatic sensitivity of roughly 0.4°C for a doubling of the CO_2 content of the air (Idso 1998).

Taking a different approach, Forest & al. (2000) defined a probability of expected outcomes by performing a large number of sensitivity runs (i.e., by varying assumptions for cloud feedback and rate of heat uptake by the deep ocean). The key statistical statement from their modeling is that there is a 95% probability that the expected global increase in surface temperature from a doubling of the CO_2 concentration in the atmosphere would range from 0.5°C to 3.3°C. It may or may not be fortuitous that the semi-empirical estimates by Lindzen and Idso fall within the acceptable range of global temperature change as deduced from the statistical work of Forest & al. (2000). The result that emerges is that current estimates from climate models of global temperature changes owing to increased concentration of atmospheric CO_2 remain highly uncertain. Forest & al. concluded that the current generation of GCMs do not cover the full range of plausible climate sensitivity.

Anthropogenic impacts upon global climate occur against a background of natural variability. There are several limitations that impede the detection of anthropogenic effects upon increased atmospheric CO_2. One is the inadequacy of climate records, which are, in general, too short to capture the full range of natural variability. For example, in the case of the interpretation of the observations of the North Atlantic oscillation, spectrum analysis reveals a climatic pattern that is randomly varying and shows little evidence for a persistent long-term trend that might be expected from an anthropogenic signal (Wunsch 1999).

Another difficulty in assessing results from models for anthropogenic effects is illustrated by the fact that models under-predict the variance of natural climatic change on decade to century time scales (Barnett & al. 1996; Stott & Tett 1998) or incorrectly predict the variance (Polyak & North 1997 [see also North 1997; Polyak 1997]; Barnett 1999) on the time scale over which the anthropogenic effect of increased CO_2 would be expected to arise. One reason that models under-predict natural climatic change is that not all causes of natural variability have been included or have been properly parameterized in the models. A few of the suspected climate *forcings* that are still poorly handled in the models are volcanic eruptions (e.g., Kondratyev 1996), stratospheric ozone variations (e.g., Haigh 1999), sulfate aerosol changes (e.g., Hansen & al. 1997), and solar particle and radiative forcing variations. The practical issue of attribution requires that such uncertainties be resolved. The radiative impact of increased atmospheric CO_2 is often treated as an anomalous and unique climatic response. For example, it has been claimed that anthropogenic effects are contained in recent and relatively short tropospheric temperature records (Santer & al. 1996). However, this claim was shown to be unsupportable in a longer record (Michaels & Knappenberger 1996; Weber 1996). In addition, the spatial limitations of such a record complicate the application of statistical methods used to infer correlations (Barnett & al. 1996; Legates & Davis 1997).

The most recent comparisons of observations and results from models fail to reveal a unique and significant change caused by increases in greenhouse gases, increases in sulfate aerosols in the atmosphere and variations in tropospheric as well as stratospheric ozone (e.g., Graf & al. 1998; Bengtsson & al. 1999). These results are consistent with analyses of circulation patterns in the northern hemisphere (Corti & al. 1999; Palmer 1999) in which the spatial patterns of anthropogenically forced climate change are indistinguishable from those of natural variability. Interpreting climate change under the perspective of such nonlinear dynamics imposes a strong requirement that a GCM must simulate natural circulation regimes and their associated variability accurately. This particular caveat is relevant because the global radiative forcing of a few watts per square meter as expected from combined anthropogenic greenhouse gases is very small compared to the energy budgets of various natural components of the climatic system and flux errors in the models' parameterizations of physical processes.

Modeling climate change is a useful approach to studying the attribution of effects of increased atmospheric CO_2. However, validation of the models is essential to placing confidence in this approach. In this regard, observations improved in precision, accuracy, and global coverage are important requirements that could aid in the early detection of a relatively weak global anthropogenic signal as well as in the improvement of the models in critical aspects.

At present, the unique attribution of climate change caused by increased concentration of atmospheric CO_2 is not possible, given the limitations of models and observed climatic parameters. The use of unverified models in making future projections of incomplete (or unknown) scenarios of climatic forcing shifts the focus from the problem of validating the models. In turn, that may lead to working with an hypothesis about the role of CO_2 in global warming that is not, but must be under the rule of science, falsifiable. Further, assessments of impacts like a rise in the sea level or altered frequencies and intensities of storms are premature. In addition, there is no clear evidence of the effect of anthropogenic CO_2 on global climate, either in surface temperature records of the last 100 years, or in tropospheric temperature records obtained from balloon radiosondes over the last 40 years, or in tropospheric temperature records obtained from MSU satellite experiments over the last 20 years. There is, however, substantial evidence for a host of beneficial effects of increased atmospheric CO_2 on the growth and development of plants.

Introduction

Increases in minor greenhouse gases are thought to cause large increases in surface and lower atmospheric temperatures. This hypothesis is based on computer climate modeling, a branch of science still in its infancy despite recent substantial strides in knowledge. To study the potential impacts upon climate of an increased concentration of greenhouse gases in the air, scientists use a variety of computer simulations, from a simple model working within only one spatial dimension to complex, three-dimensional, general circulation models that couple oceanic and atmospheric changes. The credibility of the calculations rests on the validity of the models. The only way to evaluate the models is to compare their predictions of current and past conditions to available information about the climate and look for consistencies or inconsistencies with relevant, observed climatic parameters, which ideally should be accurately measured. Although the models have tremendous potential for expanding knowledge of climate change and as such are a valuable tool for scientists, this does not guarantee accurate prediction. Hence, it is important to test the hypothesis that a significantly increased atmospheric CO_2 causes significant global climatic warming and its associated impacts.

We study two aspects of the consequences, realized and potential, of increased and increasing atmospheric CO_2. [1] One is the climatic response to increases in the concentration of atmospheric CO_2; the other is the response of plants to increases in the CO_2 content of the air. We review aspects of observed climatic parameters and compare them to predictions of the climate models. Our purpose is to assess the credibility of the models by comparing their outcomes to real-world observations. The selection of parameters we use is hardly exhaustive and we focus on parameters that highlight the weaknesses of the models, from which progress might be made. In particular, we chose global and regional surface and lower tropospheric temperatures, change at sea level (because it responds to temperature change through the interactions with sea-ice) and regional storms (e.g., Atlantic hurricanes, as a representation of the interaction between the ocean and the atmosphere). We also discuss attempts to begin folding the influences of vegetation into general circulation models, a highly complex interaction (e.g., Henderson-Sellers & al. 1996).

The second main consequence of increased atmospheric CO_2 that we discuss is the hypothesis that plant growth is enhanced under high concentrations of CO_2, that is, that elevated concentrations of atmospheric CO_2 increase growth rates of plants, biomass, and yield. This hypothesis is tested against experimental laboratory and field results; and again, rather than be exhaustive, we show a few specific examples of responses by vegetation to increased atmospheric CO_2.

Atmospheric carbon dioxide

The concentration of CO_2 in the earth's atmosphere has increased during the past century (figure 1; Keeling & Whorf 1997). Solid horizontal lines show the levels that prevailed in 1900 and 1940 (Idso 1989). The magnitude of this atmospheric increase during the 1980s was about 3 Gigatons of Carbon (Gt C) per year. Total annual anthropogenic CO_2 emissions for 1996—primarily from the use of coal, oil, natural gas and the production of cement—is estimated to be 6.52 Gt C (Marland & al. 1999).

To put these figures in perspective, consider the global carbon budget. It is estimated that the atmosphere contains 750 Gt C; the surface ocean contains 1,000 Gt C; vegetation, soils, and detritus contain 2,200 Gt C; and the intermediate and deep oceans contain 38,000 Gt C. Carbon

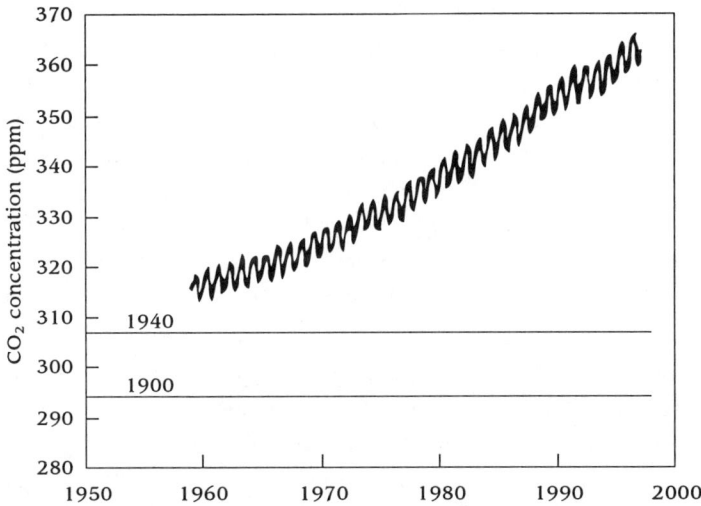

Figure 1. Atmospheric CO_2 concentration in parts per million by volume (ppm) at Mauna Loa, Hawaii (Keeling & Whorf 1997). The approximate global levels of atmospheric CO_2 in 1900 and 1940 are also displayed (Idso 1989).

shifts from one reservoir to another: each year, the surface ocean and atmosphere exchange an estimated 90 Gt C; vegetation and the atmosphere, 60 Gt C; the marine biota and the surface ocean, 50 Gt C; and the surface ocean and the intermediate and deep oceans, 100 Gt C (Schimel 1995).

So great are the magnitudes of these reservoirs, the rates of exchange between them, and the uncertainties with which these numbers are estimated, that the source of the recent rise in atmospheric carbon dioxide has not been determined with certainty (e.g., Houghton & al. 1998; Keeling & al. 1998; Peng & al. 1998; Segalstad 1998). Concentrations of CO_2 in the atmosphere are reported to have varied widely over geologic time, with peaks, according to some estimates, some twenty-fold higher than at present and troughs at approximately eighteenth-century levels (Berner 1997).

Rise in atmospheric CO_2 as a result of human activity

There is, however, a widely believed hypothesis that the rise in atmospheric carbon dioxide of 3 Gt C per year is the result of the release of carbon dioxide from human activities. This hypothesis is reasonable, since the magnitudes of human release and atmospheric rise are comparable and the atmospheric rise has occurred contemporaneously with the increase in production of CO_2 from human activities since the Industrial Revolution. Atmospheric CO_2 levels have increased substantially during the last 100 years and are expected to continue doing so. The concentration of carbon dioxide is expected to double from the pre-industrial level of 280 ppm in another 100 years or so.

However, the factors that influence the atmospheric CO_2 concentration are not fully understood. For example, the current increase in CO_2 follows a 300-year warming trend, during which surface and atmospheric temperatures have been recovering from the global chill of the Little Ice Age (see below). The observed increases in concentration of atmospheric CO_2 are of a magnitude that can, for example, be explained by oceans giving off gases naturally as temperatures rise (Dettinger & Ghil 1998; Segalstad 1998). Indeed, changes in atmospheric carbon dioxide have shown a tendency to follow rather than lead increases in global temperatures (Kuo & al. 1990; Priem 1997; Dettinger & Ghil 1998; Fischer & al. 1999; Indermühle & al. 1999). Those studies emphasize the need to understand changes in terrestrial biomass and sea-surface temperature, two important drivers of change in the concentration of atmospheric CO_2. Thus, understanding the carbon budget is a prerequisite for estimating future scenarios for concentrations of atmospheric CO_2.

Atmospheric and surface temperatures

What effect is the ongoing rise in the CO_2 content of the air having upon global temperature? In order to answer this question, one must consider the available information about temperature and its qualifications. The temperature of the earth varies naturally over a wide range, but available temperature records are spatially and temporally limited.

Reconstructed temperature from proxy records

Records going back longer than 350 years are reconstructed from proxies. A recent reconstruction of the temperature of the Northern Hemisphere from several sites yields a record going back 1000 years (Mann & al. 1999). That reconstruction is based primarily on the width and density of tree rings, which are primarily indicators of summer temperature. The record has varied over a range of no more than 1°C in the hemispheric average. There are important limitations to the interpretation of the proxy temperature. For example, Briffa & al. (1998) find width and density of tree rings have become less sensitive to recent changes in temperature (see their figure 6) over the last few decades.

Going back further means having less global information. Figure 2, for example, summarizes the temperature of the sea surface reconstructed from oxygen isotopes in the shells of *Globigerinoides ruber* in sedimentary deposits in the Sargasso Sea during the past 3,000 years (Keigwin 1996). Temperatures of the sea surface at this location have varied over a range of about 3.6°C during the past 3,000 years.

Both Mann & al.'s more widely sampled, and Keigwin's local, reconstructions display a long-term cooling trend that ends late in the nineteenth century. Two noticeable features in Keigwin's record are the Little Ice Age about 300 years ago, and the Medieval Climatic Optimum about 1000 years ago. During the Medieval Climatic Optimum, temperatures

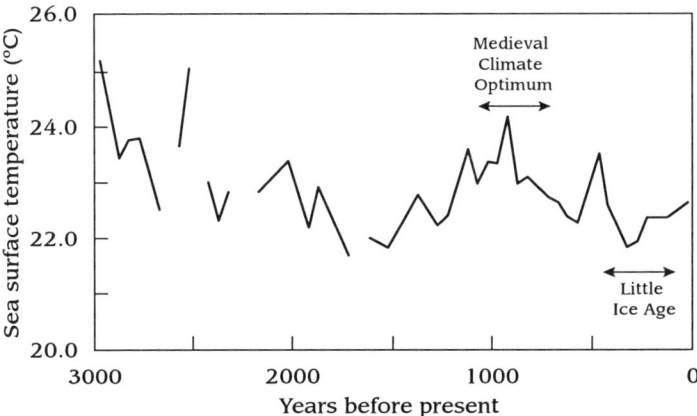

Figure 2. Surface temperatures in the Sargasso Sea (with a time resolution of about 50 years) over approximately 3,000 years (ending in 1975), as determined by oxygen isotope ratios of marine organism remains in sediment at the bottom of the sea (Keigwin 1996). The Little Ice Age and Medieval Climate Optimum are indicated.

were warm enough to allow the colonization of Greenland. The colonies were abandoned after the onset of colder temperatures, however; and, for the past 300 years, world temperatures have been gradually recovering (Lamb 1982; Grove 1996). According to Grove, the glacial record maintains a significant and coherent cooling over all continents, in agreement with the Bradley and Jones' (1993) reconstruction for the Northern Hemisphere. Thus, the evidence is that the Little Ice Age was at least a hemispheric, if not global, event. On the matter of the Medieval Climatic Optimum, several lines of evidence point to warm temperatures roughly around 1000 years BP (before present). The evidence includes montane glaciers, glacial moraines, tree growth, shell sediments and historical documentation, all indicating fairly widespread, although not strongly synchronized, warmth. For example, in China and Japan the warming ended by 900 years BP, while in Europe and North America the warming continued for two or three more centuries (Lamb 1982; Grove & Switsur 1994; Hughes & Diaz

1994; Keigwin 1996; Grove 1996). The trend to declining temperatures in the reconstructed record of the Northern Hemisphere (Mann & al. 1999) is consistent with the erosion in climate on a hemispheric scale, from 1000 years BP through the Little Ice Age about 300 years ago.

Instrumental records:
land and sea temperatures

In more recent times, instrumental records have become available. One long surface record with good quality control and coverage of a significant land area is that of the continental United States. Figure 3 shows the annual average temperature of the United States as compiled by the National Climate Data Center (Brown & Heim 1998). The upward temperature fluctuation between 1900 and 1940 is natural, because the amount of carbon dioxide added to the air from human activities was small then, and likely is a recovery from the Little Ice Age. The temperatures in

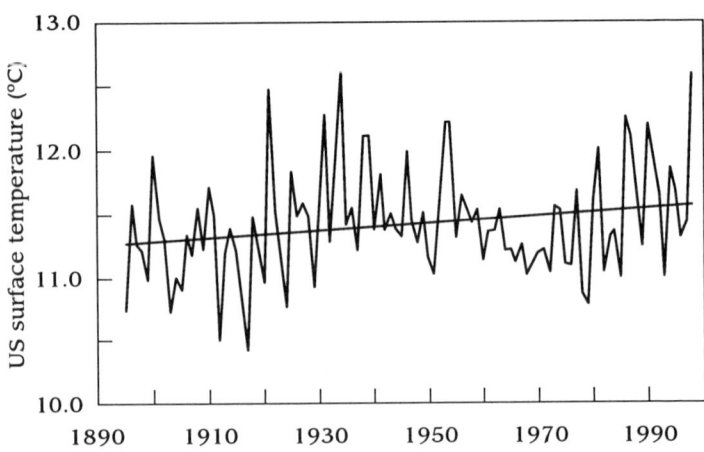

Figure 3. Annual mean surface temperatures in the continental United States between 1895 and 1998, as compiled by the National Climate Data Center (Brown & Heim 1998). The trend line for the entire data set with slope of +0.027°C per decade is indicated.

the United States show a non-significant increasing trend of +0.027°C per decade. [2]

Records of surface temperature compiled from world-wide stations by NASA-GISS (Hansen & Lebedeff 1987; Hansen & Lebedeff 1988; Hansen & al. 1996) and the Climate Research Unit (CRU) at the University of East Anglia (Parker & al. 1994; CRU 1999) are shown in figure 4. The overall rise of about 0.5°C to 0.6°C during the twentieth century is often cited in support of greenhouse global warming (e.g., Schneider 1994). However, since approximately 80% of the rise in levels of CO_2 during the twentieth century (see figure 1) occurred after the initial major rise in temperature, the increase in CO_2 cannot have caused the bulk of the past century's rise in temperature. In addition, it has been pointed out that reported increases in surface temperatures around the globe and in the northern hemisphere since the 1970s have occurred mostly during cold seasons. The winter

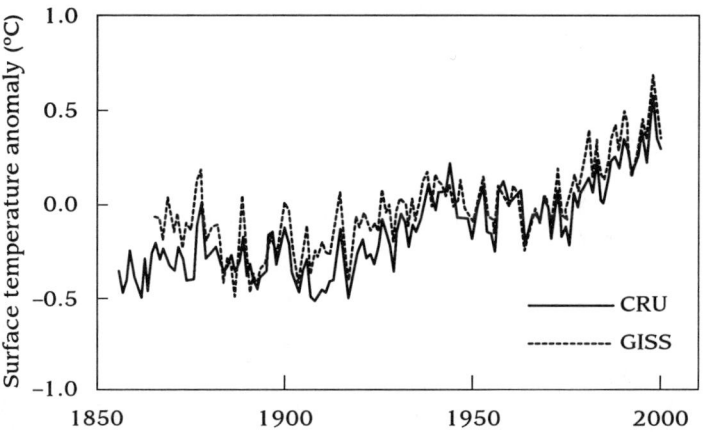

Figure 4. Annual mean global surface temperature anomalies for land and sea-surface (solid), as reconstructed by CRU (Parker & al. 1994, CRU 1999), and for land only (dashed), as estimated by NASA GISS (Hansen & Lebedeff 1987, Hansen & Lebedeff 1988, Hansen & al. 1996). An adjustment has been made for urban warming effects in both the CRU and GISS reconstructions.

warming may be interpreted as natural dynamic variability owing to anomalous atmospheric circulation. Circulation anomalies in the 1970s arise from persistently colder ocean and warmer land (COWL) surface temperatures than average (Wallace & al. 1995; Wallace & al. 1996). Could the increase in well-mixed CO_2 in the atmosphere produce the observed regional changes in the COWL pattern? Results from GCMs are inconclusive; Broccoli & al. (1998) suggest that separating the COWL pattern from the hypothesized anthropogenic CO_2 fingerprint is not straightforward (see further discussion below). Because a COWL pattern arises primarily from the contrast in the thermal inertia between the land and the sea, such an internal spatial pattern may be caused by any number of external warming influences (see also Corti & al. 1999).

Limitations of observed trends in surface temperatures: spatial coverage and uneven temporal sampling

Before interpreting other patterns of climate change, the limitations of observed spatial and temporal trends should be examined. In terms of spatial coverage, the surface records are limited because they are not truly global (Robeson 1995). The temperature trend of 0.5°C to 0.6°C over the last century has been determined with uncertainties estimated to be smaller than the magnitude of the increase (Karl & al. 1994), despite the incomplete surface coverage. But, serious uncertainty arises from two sources in uneven temporal sampling. One source of uncertainty is the unknown time of day of the observation as, for instance, in cases where only monthly data have been given (Madden & al. 1993).

A second source of error in temporal sampling is the presence of gaps in records (Stooksbury & al. 1999), which can, in turn, bias spatial coverage through the rejection of a location whose period of measurement is incomplete. For example, in the 100-year period from 1897 to 1996, Michaels

& al. (1998) found that imposing the validity requirement that data within 5° by 5° gridded spatial cells should have no more than 10 years of measurements missing (10% of the period) produced a "global" sample covering only 18.4% of the earth. This is not an optimal spatial coverage for determining a global mean. Any bias in spatial and temporal sampling can introduce significant complications and errors into the calculation of a global average.

Limitations of observed trends in surface temperatures: urban heat islands

A further uncertainty in the measurement of surface temperatures is the bias caused by urban heat islands. This bias stems from heat that is stored by, for example, the pavement of cities and raises local temperature readings above what they would be otherwise. The expansion of a city's infrastructure is in general related to the growth of the population, which is a proxy for an increasing effect of the urban heat island bias. Figure 5 shows the size of the effect of the urban heat island in temperature measurements from surface stations in California. The results from all counties and selected sites in figure 5 should be compared with the results from the East Park station—considered the best situated rural station in the state (Goodridge 1998)—which has a calculated temperature trend between 1940 and 1996 of –0.055°C per decade. The urban heat bias has also been observed elsewhere (Balling 1992; Böhm 1998). Gallo & al. (1999) caution that the designations of stations for measuring surface temperatures as urban, suburban, or rural need to be reassessed periodically because the current methodology can introduce bias in global and regional trends of surface temperatures.

The systematic error caused by the effect of urban heat islands has been extensively studied and debated and remains controversial. For example, a recent analysis of rural and urban stations finds no significant difference

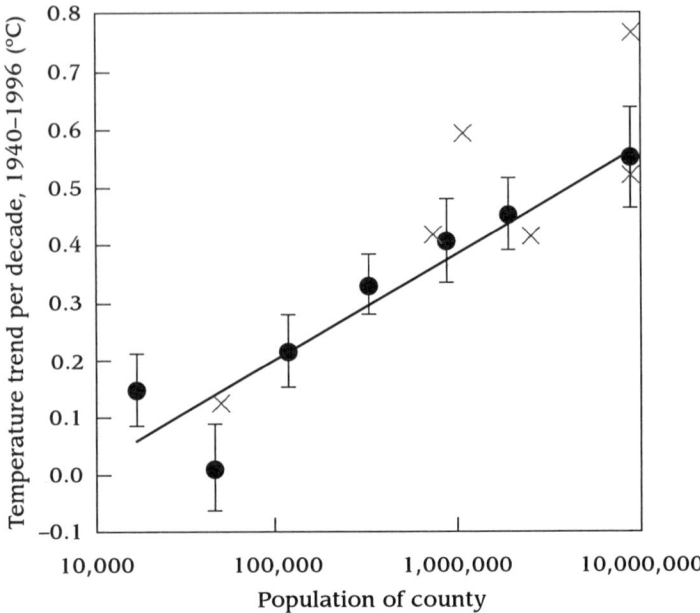

Figure 5. Surface temperature trends for the period 1940 to 1996 from 107 measuring stations in 49 California counties (Christy & Goodridge 1995, Goodridge 1996). After averaging the means of the trends in each county, counties of similar population were combined and plotted as closed circles along with the standard errors of their means. The six measuring stations in Los Angeles County were used to calculate the standard error of that county, which is plotted alone at the county population of 8.9 million. The urban heat island effect on surface measurements is evident. The straight line is a least-squares fit to the closed circles. The points marked "X" are the six unadjusted station records selected by NASA GISS for use in their estimate of global temperatures as shown in figure 4.

introduced into the calculated global trend by the urban stations until 1990 or so (Peterson & al. 1999). After 1990, the urban stations contributed a warming bias to the global average, which is not seen in the rural stations. Peterson & al. note that coverage by rural stations has fallen from over 20% of the earth's area in the 1970s to 7% in 1998, which may explain the recent difference in trends between urban and rural temperatures and may also introduce uncertainty in the global average. However, no quantitative estimate of the uncertainty can yet be made.

Tropospheric records

With the advent of the rapid increase in the concentration of atmospheric CO_2, attention has focused on the measurement of temperature trends of the last several decades. It is difficult at present to interpret reliably trends over records as short as several decades owing to uncertainties from several sources such as instruments, natural climate change caused by volcanic eruptions, or the El Nino Southern Oscillation (ENSO). In the surface record, the uncertainties are comparable to the magnitude of the trend expected from anthropogenic CO_2 (Karl & al. 1994). Tropospheric records, however, averaging temperatures from the layer of air between, roughly, the surface and an altitude of 10 kilometres, can also be considered. Although the tropospheric measurements, because they include higher altitudes and are relatively short, differ from the surface measurements, they are important in examining the effect of increases in the concentration of atmospheric CO_2. In the troposphere, temperature changes induced by greenhouse gases are expected to be at least as large as at the surface (e.g., Houghton & al. 1996).

We consider two tropospheric records: that from satellites and that from balloon platforms. Since 1979, essentially global temperature measurements of the lower troposphere have been made by means of Microwave Sounding Units (MSUs) on orbiting satellites (Spencer & al. 1990). Figure 6 shows the average global tropospheric measurements from satellites (Spencer & Christy 1990; Christy & al. 1998). The tropospheric record can be extended back to 1958 with radiosonde data (figure 7; Angell 1997, 1999). Another set of radiosonde data on tropospheric temperature by Parker & al. (1997) is also consistent with the data on tropospheric temperatures from satellites and from Angell.

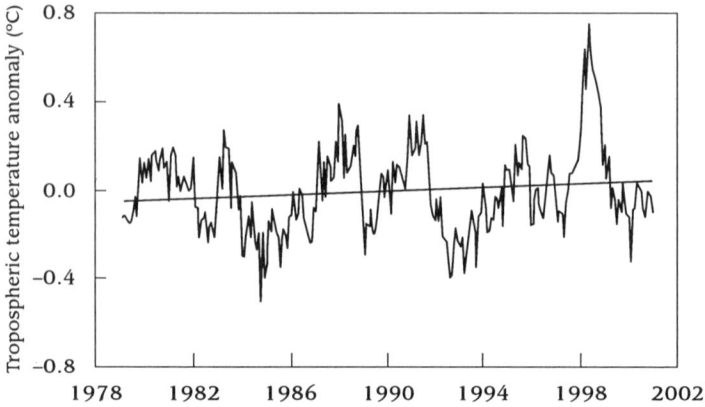

Figure 6. Satellite Microwave Sounding Unit (MSU) measurements of global lower tropospheric temperatures between latitudes 82°N and 82°S from 1979 to December 2000 (Spencer & Christy 1990, Christy & al. 1998). Temperatures are monthly averages and the linear trend for 1979 to May 1999 is shown. The slope of this line is +0.044 °C per decade.

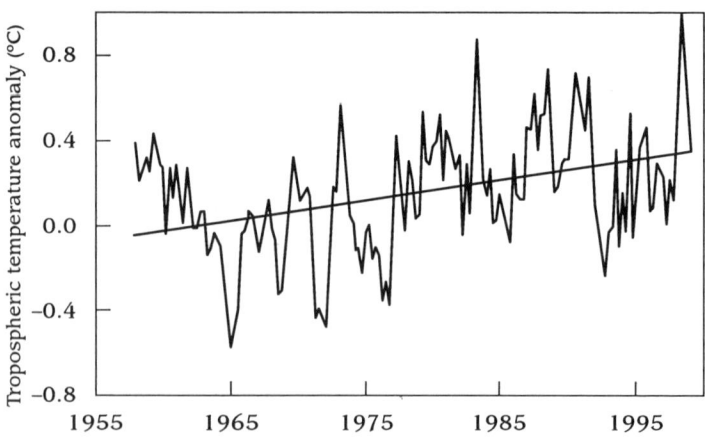

Figure 7. Radiosonde balloon station measurements of lower tropospheric temperatures at 63 stations between latitudes 90°N and 90°S from 1958 to February 1999 (Angell 1997, 1999). Temperatures are three-month averages and the linear trend for 1958 to February 1999 is shown. The slope is +0.095°C per decade.

The agreement of the independent sets of data between 1979 and 1998 verifies their precision. Further agreement between records from balloons and from satellites has been shown rigorously by extensive analysis (Spencer & Christy 1992; Christy 1995).

An analysis of the satellite record (Wentz & Schabel 1998) had pointed out a potential error in the calculated trend in the MSU measurements of the tropospheric temperature. The authors reported that decay of the orbits of the individual satellites contributed to uncertainty that had been previously ignored in the MSU records. However, the authors' estimate of the uncertainty turned out to be exaggerated. The effect of orbital decay, when properly computed, and the additional effect of drift in the satellites' orbits have now been applied to the MSU data (Christy & al. 1998; Christy & al. 2000). The corrected data are those plotted in figure 6.

Tropospheric temperatures have shown a global trend of +0.044°C per decade (MSU, 1979 to December 2000) or +0.018°C per decade (radiosonde, 1979 to February 1999). Going back further, the trend in the tropospheric temperatures measured by radiosonde is +0.095°C per decade (1958 to February 1999). The physical significance of the tropospheric trends is difficult to assess. These periods are relatively short intervals over which to interpret a trend because of large interannual variability (caused, for example, by the effects of ENSO and volcanic eruptions). But Christy & McNider (1994, and updates in Christy 1997) have shown that after the crude removal of the effects of ENSO and volcanic eruptions, the adjusted 20-year trend of tropospheric global temperatures measured by MSUs can be only marginally consistent with the warming rates of 0.08°C to 0.30°C per decade projected by GCM simulations that included effects owing to CO_2 (only) and tropospheric sulfate aerosol (direct effect only) over the last 20 years (Houghton & al. 1996: 438).

The trends in tropospheric temperatures can be compared to the trend in the CRU's record of surface temperature of +0.11°C per decade (1958–1998) and +0.19°C per decade (1979–1998). The surface trends are apparently significant but are not consistent with the tropospheric trends. No clear resolution of this important discrepancy is at hand. Nevertheless, in addition to uncertainties in the surface measurements, there are physical reasons for the differences expected between the surface and tropospheric temperature trends that one finds, for example, over the equatorial oceans (Trenberth & al. 1992, Christy 1995). (See discussion below on the complications in attributing causes to recent climate change.) New attempts after Houghton & al. (1996), like Bengtsson & al. (1999), have further highlighted the inconsistency between the observed trends in surface and troposphere temperatures and the simulated GCM trends that try to include forcing factors like anthropogenic greenhouse gases, tropospheric sulfate aerosols (both the direct and indirect effects), stratospheric aerosols from Mount Pinatubo, as well as tropospheric and stratospheric ozones.

Stratospheric records

Another useful record is that of stratospheric temperatures, which have been measured by MSU from 1979 and by balloon—although the area and altitude covered have been more restricted than those covered by the satellite record—from 1958. The lower stratosphere (about 100 millibars) has shown a significant cooling trend of about 0.6°C to 0.7°C per decade since 1979 (see e.g., Angell 1997, Parker & al. 1997, Simmons & at. 1999). Qualitatively, the cooling trend is apparently consistent with the expectation of cooling caused by increased radiative emission as a consequence of increased concentration of CO_2 in the stratosphere. However, determining the potential CO_2 component of stratospheric cooling is difficult because the effects of volcanic aerosols, changes in stratospheric ozone, and changes in solar ultraviolet forcing can be significant. These three effects can change with time and cannot be removed with precision. The technical difficulty of attributing the causes of lower stratospheric cooling, as well as variations of the surface and tropospheric temperatures, is discussed further below.

Modeled effects of increased atmospheric CO_2

Incoming broadband solar radiation is, on average, approximately balanced by the outgoing thermal radiation of the Earth. The major greenhouse gas, H_2O (in vapor or condensed forms like clouds), and minor greenhouse gases like CO_2, CH_4 and N_2O act to regulate this radiational balance by absorbing and re-emitting large portions of the outgoing, infrared terrestrial radiation. Introduction of additional CO_2 into the atmosphere can be considered as an effective increase in radiative energy input, one type of forcing to the climate system. The heat is then redistributed, both vertically and horizontally, by various processes involving motions in the atmosphere and ocean.

When CO_2 increases the capacity for trapping infrared energy in the atmosphere, how does the atmosphere respond? The radiative contribution of doubling atmospheric CO_2 is not large and the total response depends on many damping or amplifying processes, called feedback mechanisms. This is the key issue. Many of the physical processes are only understood in the most rudimentary fashion and are variously parameterized, yielding different computed responses among the models for doubling the concentration of atmospheric carbon dioxide. Thus, the computer-generated climate models have substantial uncertainties (Mason 1995). Without experimental validation of the models, the calculation of the climatic response to increased anthropogenic atmospheric CO_2 is not reliable. Also needed is a reliable calculation of the natural variability of the climate. We discuss six important areas in climate modeling.

Energy flux errors
A climate model follows the flow of energy, or energy flux, through the components of the climate system. Nearly

all models have substantial errors in the calculation of the energy flux and introduce artificial flux adjustments to compensate for these errors. Systematic heat-flux errors between the surface and the ocean of up to 100 Watts per square meter are locally introduced into some calculations (Glecker & al. 1995; Murphy 1995; Glecker & Weare 1997). One important consequence of such flux adjustments is to damp low-frequency variability in the simulation of a climate state through excessive over-stabilization (Palmer 1999). Another critical consequence of the artificial flux tuning is to introduce systematic biases in the model's estimates for important parameters of the climate system like the annual mean and annual cycle amplitude of the equator-to-pole temperature gradient and the ocean-land surface temperature contrast (Jain & al. 1999). Several coupled ocean-atmosphere models attempt to avoid flux adjustments. However, those models still show substantial climate drift and bias (Cai & Gordon 1999; Yu & Mechoso 1999).

In addition, results from the cooperative CAGEX, [3] CEBA, [4] as well as the latest ARESE [5] experiments (e.g., Wild & al. 1995; Charlock & Alberta 1996; Valero & al. 1997a; Zender & al. 1997) show that there is atmospheric absorption unaccounted for by the present atmospheric radiation codes generally used in GCMs. Those observations suggest that the missing energy flux, when averaged over the whole globe, amounts to 25 Watts per square meter (Cess & al. 1995; Li & al. 1997) or 17–20 Watts per square meter (Zhang & al. 1998) or 10–20 Watts per square meter (Wild & al. 1998). The missing energy flux, interpreted as excess cloud absorption, occurs in both visible (224–680 nm) and near-infrared (680–3300 nm) wavelengths, while an excess absorption around 500 nm (with 10 nm bandwidth) can be ruled out (Valero & al. 1997b; Cess & al. 1999). The flux uncertainties are large compared to the expected forcing from doubling CO_2, about 4 Watts per square metre globally.

Water vapor feedback

A second important factor in climate modeling is the under-standing of water-vapor feedback. The underlying process starts with increasing temperature that increases concen-tration of atmospheric water vapor. The models assume that the water vapor is then distributed, especially to the middle and upper troposphere, in such a way as to increase water-vapor content globally. Consequently, the enhanced water vapor would amplify the warming caused by increased CO_2 alone. This is the dominant gain in the models for amplify-ing the effect of CO_2 increases. This mechanism has been studied theoretically and observationally. Some evidence supports a positive water-vapor feedback (Liao & Rind 1997; Soden 1997; Inamdar & Ramanathan 1998). However, the model parameterization of the mechanism has been crit-icized (Renno & al. 1994; Spencer & Braswell 1997). For example, the interannual variations of water vapor and tem-perature in the tropical troposphere have been shown to be too strongly coupled in a GCM when compared to observed relationships (Sun & Held 1996). A comparison of observed decadal mean tropospheric precipitable water over North America, the Pacific basin and the globe with results from 28 GCMs revealed that simulated values are less moist than the real atmosphere for all three cases (Gaffen & al. 1997). Tropospheric moisture and convective transport processes are both specific to the spatial and temporal scale on which they occur (Hu & Liu 1998; Yang & Tung 1998). Without adequate observations, it is difficult to determine the correct parameterization of the effect of the CO_2-induced water-vapor feedback. Limited observations of precipitable water have been obtained in the tropics (30°N–30°S) and yield an indication of widespread drying of the upper tropo-sphere between 1979 and 1995 (Schroeder & McGuirk 1998a; see also the exchange between Ross & Gaffen 1998 and Schroeder & McGuirk 1998b).

Cloud forcing and feedback

A third factor limiting the performance of the models is uncertainty related to cloud forcing. Necessary observations of cloud properties are incomplete although current intensive programs are progressing (Rossow & Cairns 1995; Hahn & al. 1996; Weare 1999; Wylie & Menzel 1999). In general, GCMs over-predict the coverage (cloudiness) of high clouds by a factor as large as 2 to 5 (Weare & AMIP Modeling Groups 1996). For low clouds, models present a global average coverage that is 10% to 20% less than observed. Spatially, the modeled cloud distribution is also incorrect (Weare & AMIP Modeling Groups 1996; Cess & al. 1997). The magnitude of such systematic errors in cloud parameterization is not negligible. Additional physical processes like the feedback from changes in the production of cirrus clouds as a result of variable detrainment temperatures (Chou & Neelin 1999) and the impact of clouds on the spectral distribution of incident irradiance (Siegel & al. 1999) should also be included. Therefore, the parameterization of radiative, latent and convective effects of cloud forcing needs further improvements (Fowler & Randall 1999; Rotstayn 1999; Senior 1999; Yao & Del Genio 1999).

Interaction between ocean and atmosphere

A fourth critical area of uncertainty concerns the parameterization of the interaction between ocean and atmosphere. The physics of the interaction between air and sea is actively being studied, especially over the tropical oceans, with observations *in situ* and by satellite of heat, momentum and freshwater fluxes. Observations of such parameters should improve our understanding of air-to -sea coupling (e.g., Godfrey & al. 1998).

As just one example of the inadequacies of current models in this area, consider hurricanes and storms. Modeling storms is complex: for instance, storms occur on spatial scales of roughly tens of kilometres, below the resolution (hundreds of kilometres) of most models (e.g., Hendersen-Sellers & al. 1998). As for results of roughly doubling the concentration of atmospheric CO_2, one model calls for hurricane wind speeds to increase 3 to 7 meters per second and for central surface pressures to drop by 7 to 20 millibars over the western Pacific (Knutson & al. 1998). Another model predicts increases in precipitation extremes almost everywhere under the scenario of a doubling of the concentration of atmospheric CO_2 (Zwiers & Kharin 1998). Both studies admit to significant imprecision arising from uncertainties in the sea-to-air interaction and other model deficiencies like inadequate spatial resolution.

In testing these various predictions, storm assessment is possible in regions of the Atlantic where data go back 100 years or so. Figure 8 shows the number of severe tropical Atlantic hurricanes per year and also the maximum wind intensities of those hurricanes (Landsea & al. 1996). Both of these parameters have been decreasing with time. Another study that focused on a subset of tropical Atlantic hurricanes, those from the Gulf of Mexico making land-fall in the United States, also showed no sign of increasing hurricane frequency or intensity over the period 1886 to 1995 (Bove & al. 1998). Likewise, in regions of the northern Atlantic, there is interdecadal variation in the storm index but, over more than a century from 1881 to 1995, no increasing or decreasing trend in storm roughness (WASA Group 1998).

Feedback from sea ice and snow

A fifth important process for models to simulate is the response of sea ice and snow to a forcing like doubled concentration of carbon dioxide in the air (Randall & al. 1998). GCM results under-predict the variance of sea-ice thickness in the Arctic on decadal to century time scales (Battisti & al. 1997). Zhang & al. (1998) emphasized the importance of including realistic surface fluxes and modeling of convec-

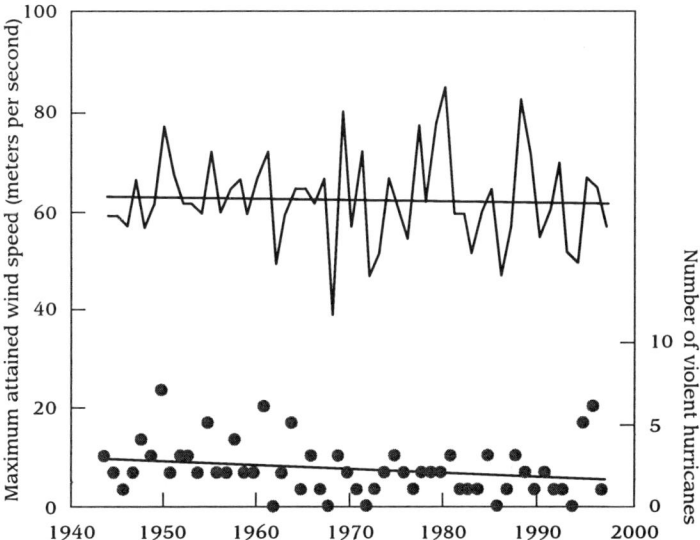

Figure 8. Annual numbers of violent Atlantic Ocean hurricanes (right scale) and their maximum attained wind speeds (left scale) (Landsea & al. 1996). Slopes of the trend lines are –0.25 hurricanes per decade and –0.33 meters per second maximum attained wind speed per decade.

tive overturning and vertical advection both in the Arctic and adjacent oceans to correct the overly warm intermediate layers in the Arctic ocean and excessive heat influx into the Fram Strait predicted by models. An analysis of results from 27 GCMs reveals that most of the models display less than half the interannual variance of snow extent; furthermore, the models underestimate snow extent in some areas and overestimate it in others (Frei & Robinson 1998).

A related consequence of feedback from sea ice and snow is a change in sea level, a measurement that can be used to assess the gross state of the model results and its parameterization of the feedback from sea ice and snow. Figure 9 shows satellite measurements of global variations of sea level between 1993 and 1996 (Nerem & al. 1997). The reported current global rate of rise amounts to about +2 mm per year (Douglas 1995; Nerem & al. 1997). The trends in rise and fall of sea level in various regions have a wide range of about 100 mm per year with most of the globe showing downward trends except near the eastern equatorial Pacific (Douglas 1995; Nerem & al. 1997; Leuliette & Wahr 1999). Historical records show no acceleration in the rise of the sea level in the twentieth century (Douglas 1992). Such observations seem inconsistent with even the modest predictions by models of a rise in sea level during the twentieth-century as a result of oceanic thermal expansion. Predictions of a CO_2-induced rise in the sea level will likely remain uncertain for some time because most

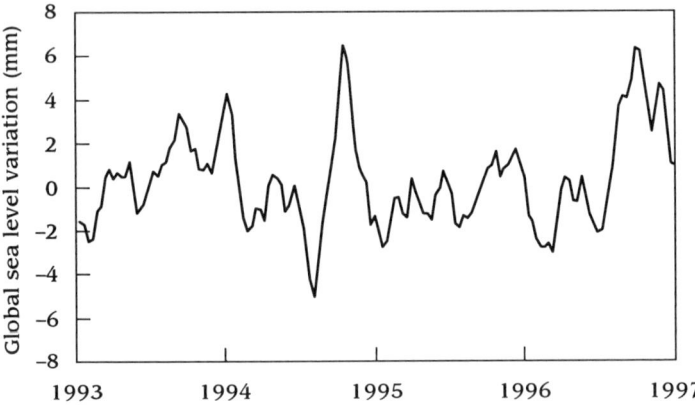

Figure 9. Global sea level variations from the Topex/Poseidon (T/P) satellite altimeters for 1993 to 1996 (as in figure 1 of Nerem & al. 1997; the curve smoothed by a 60-day boxcar filter, plotted for each 10-day cycle of the T/P altimetry). No trend is calculated here, but according to Nerem & al. (1997), the plotted T/P instrumental time series gives a 'secular' rate of change of about -0.2 mm per year, after the removal of annual and semi-annual variations (p. 1332 of Nerem & al. 1997, see e.g., Leuliette & Wahr 1999 for updates on the analysis of spatial and temporal variability of T/P sea surface height data). However, it has been reported that 50-year tide gauge measurements give +1.8 mm per year (Douglass 1995). A correction of +2.3 mm per year was added to the satellite data based on comparison to selected tide gauges to get a value of about +2.1 mm per year (Nerem & al. 1997).

of the factors affecting sea-level change, including not only the sea-ice feedback but also vertical land motion, are not well understood and are difficult to model (Douglas 1995; Gornitz 1995; Peltier 1996; Conrad & Hager 1997).

Feedback from biosphere, atmosphere and ocean

A sixth important set of processes to be considered in climate modeling has to do with feedback from biosphere, atmosphere and ocean (Idso 1989). These processes are difficult to incorporate in models, yet progress is being made (e.g., Henderson-Sellers & al. 1996; Varejao-Silva & al. 1998). A small positive feedback effect on global temperature sensitivity has been found in a GCM that includes some effects of plant photosynthesis and soil thermodynamics (Cox & al. 1999). For latitudes above 45°N, Levis & al. (1999) have found substantial spring and summer warming and winter cooling effects (primarily through alteration of *surface albedo*) when vegetation feedbacks are incorporated into a GCM under a doubled CO_2 scenario. Soil moisture changes appreciably when biospheric processes are included; and including them should reduce systematic errors in the simulation of regional climate. The most important aspect of biospheric feedback is perhaps its relevance to the carbon budget of the climate system. Understanding this feedback holds the promise of an internally consistent description of the relationship of CO_2 to climate change.

Effects of increased CO_2 on plants

Plant life provides a sink for atmospheric CO_2. Using current knowledge about the increased growth rates of plants and assuming a doubling of CO_2 release as compared to current emissions, it has been estimated that atmospheric CO_2 levels will rise by about 300 ppm before leveling off (Idso 1991). At that level, CO_2 absorption by increased terrestrial biomass may be able to absorb about 10 Gt C per year.

Figures 10 to 13 show examples of experimentally measured increases in the growth of plants. These examples are representative of a very large research literature on this subject (Kimball 1983; Cure & Acock 1986; Mortensen 1987; Drake & Leadley 1991; Lawlor & Mitchell 1991; Gifford 1992; Poorter 1993). Since plant response to fertilization by CO_2 is nearly linear with respect to the concentration of CO_2 over a range of a few hundred ppm, as seen, for example, in figures 10 and 13, it is easy to normalize experimental measurements at different levels of CO_2 enrichment. This has been done in figures 14a and 14b in order to illustrate some CO_2 growth enhancements calculated for the atmospheric increase of about 80 ppm that has already taken place, and that expected from a projected total increase of 320 ppm.

Figure 10 summarizes the increased growth rates of young pine seedlings at four levels of CO_2. Again, the response is remarkable, with an increase of 300 ppm more than tripling the rate of growth (Idso & Kimball 1994). Figure 11 shows the effect of CO_2 fertilization on sour orange trees (Idso & Kimball 1991, 1997). During the early years of growth, the bark, limbs, and fine roots of sour orange trees growing in an atmosphere with 700 ppm of CO_2 exhibited rates of growth more than 170% greater than those at 400 ppm. As the trees matured, this CO_2-induced enhancement dropped to about 100%. Meanwhile, orange production was 127% higher for the 700 ppm trees.

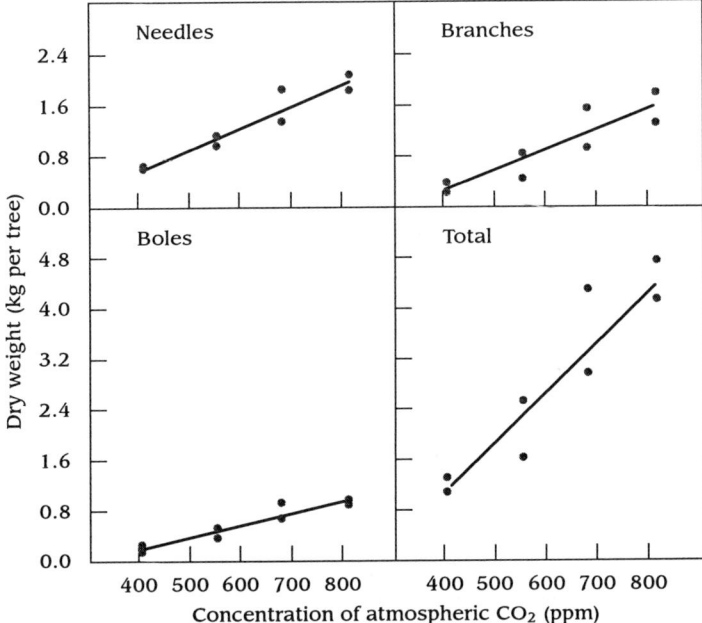

Figure 10. Young Eldarica pine trees were grown for 23 months under four CO_2 concentrations and then cut down and weighed. Each point represents an individual tree (Idso & Kimball 1994). Weights of tree parts are as indicated.

Trees respond to CO_2 fertilization more strongly than do most other plants but all plants respond to some extent. Figure 12 shows the response of wheat grown under wet conditions and when the wheat was stressed by lack of water. These were open-field experiments. Wheat was grown in the usual way but the atmospheric concentration of CO_2 in circular sections of the fields was increased by means of arrays of computer-controlled equipment that released CO_2 into the air to hold the levels as specified.

While the results illustrated in figures 10 to 12 are remarkable, they are typical of those reported in a very large number of studies of the effects of CO_2 concentration upon the growth rates of plants.

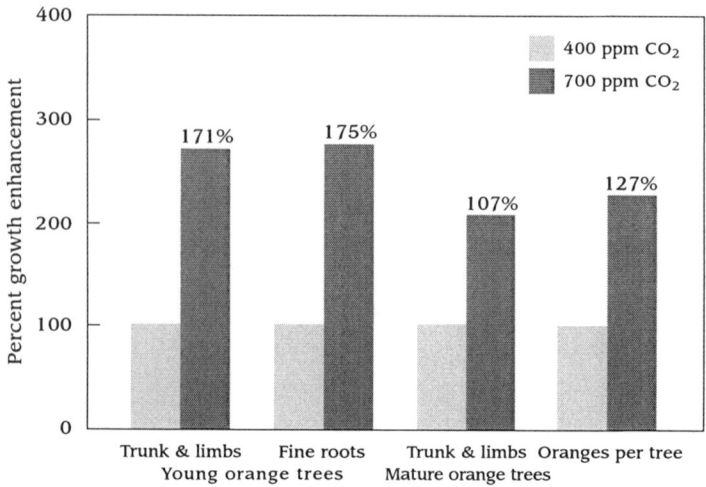

Figure 11. Relative trunk and limb volumes and fine root biomass of young sour orange trees; and trunk and limb volumes and numbers of oranges produced per mature sour orange tree per year at 400 ppm CO_2 (light bars) and 700 ppm CO_2 (dark bars) (Idso & Kimball 1991, Idso & Kimball 1997). The 400 ppm values were normalized to 100. The trees were planted in 1987 as one-year-old seedlings. Young trunk and limb volumes and fine root biomass were measured in 1990. Mature trunk and limb volumes are averages for 1991 to 1996. Orange numbers are averages for 1993 to 1997.

Figure 13 summarizes 279 similar experiments in which plants of various types were raised under CO_2-enhanced conditions. Plants under stress from less-than-ideal conditions—a common occurrence in nature—respond more to CO_2 fertilization. The selections of species shown in figure 13 were biased toward plants that respond less to CO_2 fertilization than does the mixture actually covering the Earth, so figure 13 underestimates the effects of global CO_2 enhancement.

Figures 14a and 14b summarize the enhancements of wheat, orange trees, and young pine trees shown in Figures 12, 11 and 10 with two idealized atmospheric CO_2 increases—that which has occurred since 1800 and is believed to

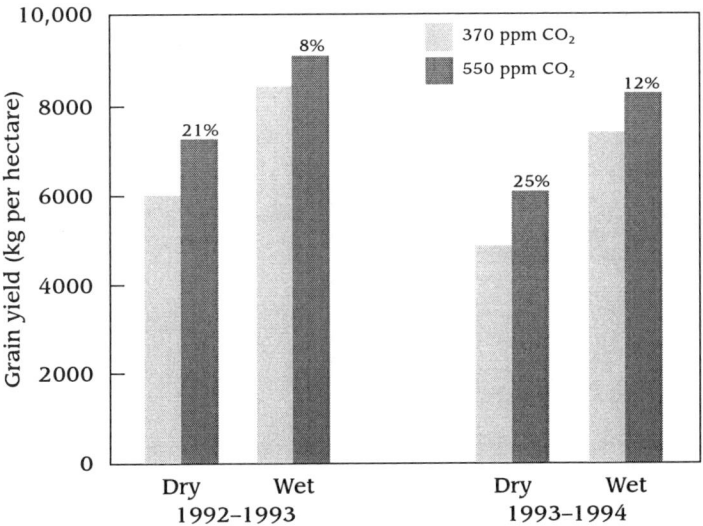

Figure 12. Grain yields from wheat grown under well watered and poorly watered conditions in open field experiments (Kimball & al. 1995, Pinter & al. 1996). Average CO_2-induced increases for the two years were 10% for wet and 23% for dry conditions.

be the result of the Industrial Revolution, and that which is projected for the next two centuries.

Carbon dioxide is not a pollutant; it is essential to life. Based on extensive evidence from agricultural research on enhanced carbon dioxide environments both in the field and in laboratories, increases in carbon dioxide should cause many plants to grow more vigorously and quickly. The reason is that most plants evolved under, and so are better adapted to, concentrations of atmospheric carbon dioxide higher than those found at present. In experiments doubling the air's carbon dioxide content, the productivity of most herbaceous plants rises 30% to 50%, while the growth of woody plants rises more so. The impacts of enhanced plant growth and related soil changes may even provide a strong quenching effect upon warming from carbon dioxide. The vegetation feedbacks as a result of carbon

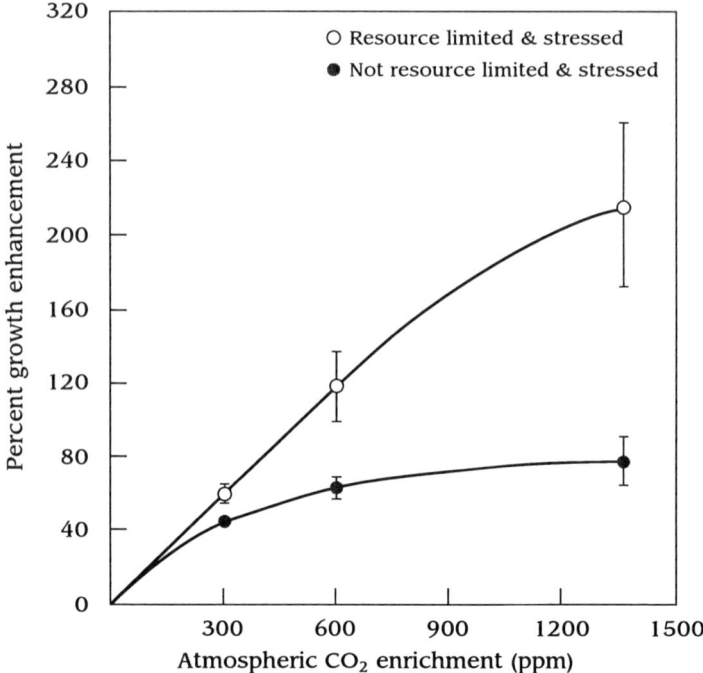

Figure 13. Summary data from 279 published experiments in which plants of all types were grown under paired stressed (open circles) and unstressed (closed circles) conditions (Idso & Idso 1994). There were 208, 50, and 21 sets of data at 300, 600, and an average of about 1350 ppm CO$_2$, respectively. The plant mixture in the 279 studies was slightly biased toward plant types that respond less to CO$_2$ fertilization than does the actual global mixture and therefore underestimates the expected global response. CO$_2$ enrichment also allows plants to grow in drier regions, further increasing the expected global response.

dioxide fertilization have yet to be correctly incorporated in the climate simulations.

Partly as a result of elevated carbon dioxide in the air and more efficient agricultural practices, the United States has experienced in recent decades enhanced growth in vegetation. The acceleration of plant growth is of a magnitude that the United States, despite its energy use and resultant prosperity, may not be a net emitter of carbon.

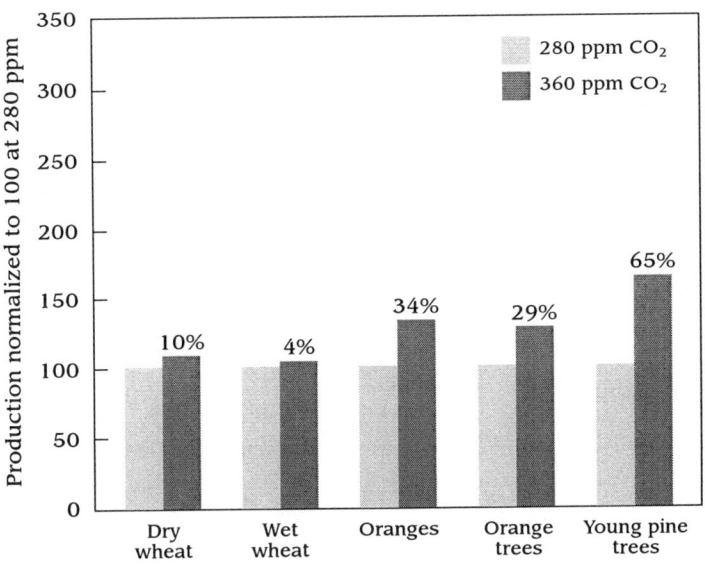

Figure 14a. Calculated growth rate enhancement of wheat, young orange trees, and very young pine trees already taking place as a result of atmospheric enrichment by CO_2 during the past two centuries. These values apply to pine trees during their first two years of growth and orange trees during their 4th through 10th years of growth. As is shown in figure 11, the effect of increased CO_2 gradually diminishes with tree age, so these values should not be interpreted as applicable over the entire tree lifespan. There are no longer-running controlled CO_2 tree experiments.

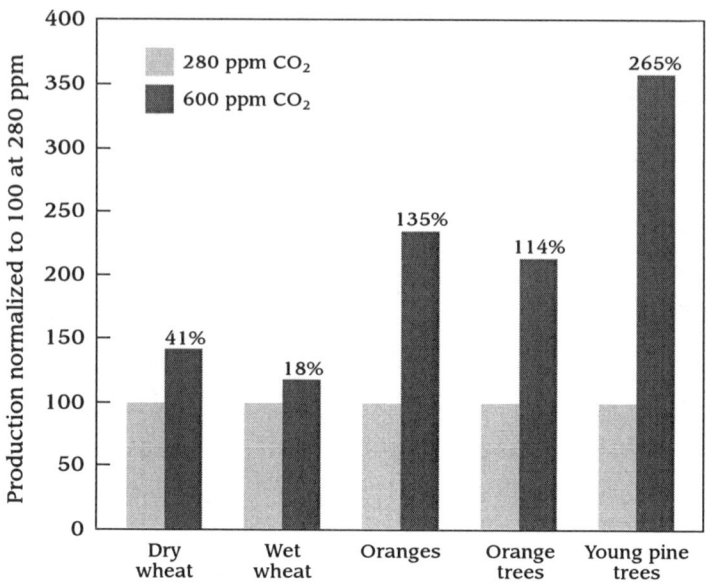

Figure 14b. Calculated growth rate enhancement of wheat, young orange trees, and very young pine trees expected to take place as a result of atmospheric enrichment by CO_2 to a level of 600 ppm.

Conclusion

There is no doubt about the improvement of the human condition through unfettered access to energy. Use of energy may also produce local unwanted pollutants as a by-product. Those sources of true environmental pollution may be tolerated or mitigated, based on rational considerations of the risks of pollutants and benefits of energy use. But, in the case of recent fears of anthropogenic carbon dioxide, science indicates that, at most, a little warming and, certainly, better plant growth will result from the projected future increase of carbon dioxide content in the air. An optimal warming and enhanced plant growth should be of great benefit to mankind and the environment.

Notes

1 For the purposes of discussion in this paper, we will use carbon dioxide as a surrogate for itself and the other minor greenhouse gases (GHGs) that have been associated with the global-warming hypothesis. Greenhouse warming models usually assume that the input of all the minor greenhouse gases, including carbon dioxide, produces an effect roughly twice that of CO_2 alone.

2 In several places in this review, we report linear least-squares calculated trend lines for measurement compilations of temperatures, sea levels, and storm activity. Further analyses for these compilations are reported in the referenced literature. Error estimations for the trend values are beyond the scope of this review. Systematic errors like temporal or spatial limitations of the data introduce substantial complication and uncertainty. The calculated trend values are, however, helpful in understanding the graphically displayed observational results.

3 CAGEX-CERES/ARM/GEWEX = NASA's Clouds and the Earth's Radiation Energy System (CERES); Department of Energy's Atmospheric Radiation Measurement (ARM); World Climate Research Program's Global Energy and Water Cycle Experiment (GEWEX).

4 World Climate Research Program's Global Energy Balance Archive (GEBA).

5 Departments of Energy and Defense and NASA's Atmospheric Radiation Measurements Enhanced Shortwave Experiment (ARESE).

Glossary

Balloon Radiosonde Temperature   Direct *in-situ* measurements of temperature in the vertical air column using thermometer carried by weather balloons. Balloon radiosonde temperature records are available for both the troposphere and stratosphere (usually information on humidity, wind and pressure are also recorded).

ENSO   The El Nino Southern Oscillation is a warming, then cooling of the equatorial Pacific ocean and lower atmosphere that recurs roughly every 3 to 7 years. ENSO has significant global impacts.

Forcing and Feedback   Forcing is an input to the climate system that leads to climatic responses and related effects. Feedback is an interaction among components and processes within the climate system that can either amplify (positive feedback) or weaken (negative feedback) an initial change or forcing.

General Circulation Model (GCM)   A model of the climate system that accounts for the budgets of energy (related to thermodynamical changes) and momentum (related to dynamic motions) in a self-consistent manner. The most sophisticated version of a General Circulation Model covers the full domain of the ocean and atmosphere and thus has to keep track of large number of climate variables and their interactions.

Gridding of data   This is a process in attempting to simplify the complex local information by an optimal choice of data averaging over a larger areal extent. For surface temperature data, it is estimated that two neighboring points are not significantly correlated (and thus can be considered a distinct local area) if the distance between them is over 1000 to 2000 km.

Major and minor greenhouse gases   Major greenhouse gas is water vapor that has a concentration close to 1% of the air by volume while minor greenhouse gases include elements like carbon dioxide, methane, nitrous oxide, ozone, chlorofluorocarbons and even the trifluoromethyl sulfur pentafluoride (or chemical compound with the label SF_5CF_3) that are measurable in parts per million (or 0.0001%) down to the minute amount of parts per trillion (10 to the power of 12) by volume. All greenhouse gases share the similar property that they are efficient absorbers and emitters of infrared radiation.

Microwave sounding unit (MSU)   An instrument designed to collect microwave light emissions from uniformly distributed molecules of oxygen in the global atmosphere. The MSU can provide precise measurements of the atmospheric temperature of the entire earth because (1) oxygen makes up 21% of our air and its concentration is very stable in time and space; (2) the microwave radiation emitted by the oxygen molecules can pass through the earth's clouds, including the high-altitude cirrus clouds, without much obstruction; and (3) the MSU is very well calibrated against external standards.

Parameterization   This is a practical procedure of simplifying the representation of complex physical processes in a climate system. Parameterization can be considered as an educational assumption but it is important to recognize that in the process of parameterization reality has also been greatly simplified. A good example is formation of clouds: cloudiness in a GCM is usually presented as a function of simple variables like humidity, temperature and motion within a grid box (which is typically of few hundred kilometers in size!) of the climate model and, thus, does not contain any information on details of cloud microphysics like aggregation and growth of water droplets to form real clouds.

Proxy temperature records   Indirect determination of temperature from various climate indicators like widths of tree-growth rings, isotope abundance in ice core, isotope abundance in lake and deep-sea sediments. It is important to note that no indicators give precise information on temperature changes and all proxies must clearly demonstrate high-level of direct association to instrumental data to be reliable. Some proxies can provide global information while most give local conditions.

Stratospheric temperature   Direct remote-sensing measurements of air temperature in the layer ranging from 10 km to about 50 km from microwave emission from oxygen molecules which is linked to temperature. The stratosphere contains about 10% of the air mass of Earth atmosphere.

Sulfate aerosols   Tiny liquid or solid particulates in the air made from dust particles, sulfur dioxide gas and water.

Surface albedo   A gross term used to describe the reflection of the sun's radiation from the earth's surface. The amount of reflected sunlight depends strongly on the type of surface, vegetation cover, snow and ice covers, and so on. If the surface albedo is high, more radiation will be reflected back into the atmosphere and space. If the albedo is low, more radiation is absorbed by the earth's surface.

Tropospheric temperature   Direct remote-sensing measurements of air temperature in the layer ranging from the surface to about 10 km; measures microwave emission from oxygen molecules, which is linked to temperature. Troposphere contains about 90% of the air mass of the earth's atmosphere.

Literature cited

Angell JK (1997) Annual and seasonal global temperature anomalies in the troposphere and low stratosphere, 1958-1996. In trends online: a compendium of data on global change, Oak Ridge National Laboratory (http://cdiac.esd.ornl.gov/ftp/ndp008r4/; additional updates in ftp://gus.arlhq.noaa.gov/pub/climate/angell_anoms/temperature/layer-mean/L850-300.txt)

Angell JK (1999) private communication (with data of Angell (1997) updated to February 1999)

Balling Jr. RC (1992) The heated debate. Pacific Research Institute, San Francisco

Barnett TP (1999) Comparison of near-surface air temperature variability in 11 coupled global climate models. J Clim 12: 511–518

Barnett TP & al. (1996) Estimates of low frequency natural variability in near-surface air temperature. Holocene 6: 255–263

Battisti DS, Bitz CM, Moritz RE (1997) Do general circulation models underestimate the natural variability in the Arctic climate? J Clim 10: 1909–1920

Bengtsson L, Roeckner E, Stendel M (1999) Why is the global warming proceeding much slower than expected? J Geophys Res 104: 3865–3876

Berner RA (1997) The rise of plants and their effect on weathering and atmospheric CO_2. Science 276: 544–545

Böhm R (1998) Urban bias in temperature time series—a case study for the city of Vienna, Austria. *Climatic Change* 38: 113–128

Bove, MC, Zierden DF, O'Brien JJ (1998) Are Gulf landfalling hurricanes getting stronger? *Bull Amer Meteorol Soc* 79: 1327–1328

Bradley RS, Jones PD (1993) "Little Ice Age" summer temperature variations: Their nature and relevance to recent global warming trends. *Holocene* 3: 367–376

Briffa KR & al. (1998) Trees tell of past climates: But are they speaking less clearly today? *Phil Trans R Soc Lond B* 353: 65–73

Broccoli AJ, Lau N-C, Nath MJ (1998) The cold ocean-warm land pattern: model simulation and relevance to climate change detection. *J Clim* 11: 2743–2763

Brown WO, Heim RR (1998) National Climate Data Center, *Climate Variation Bulletin* 10, 12, *Historical Climatology Series* (http://www.ncdc.noaa.gov/ol/ documentlibrary/ cvb.html/)

Cai WJ, Gordon HB (1999) Southern high-latitude ocean climate drift in a coupled model. *J Clim* 12: 132–146

Cess RD & al. (1995) Absorption of solar radiation by clouds: observations versus models. *Science* 267: 496–499

Cess RD & al. (1997) Comparison of the seasonal change in cloud-radiative forcing from atmospheric general circulation models and satellite observations. *J Geophys Res* 102: 16593–16603

Cess RD & al. (1999) Absorption of solar radiation by the cloudy atmosphere: Further interpretations of collocated aircraft measurements. *J Geophys Res* 104: 2059–2066

Charlock TP, Alberta TL (1996) The CERES/ARM/GEWEX experiment for the retrieval of radiative fluxes with satellite data. *Bull Amer Meteorol Soc* 77: 2673–2683

Chou C, Neelin JD (1999) Cirrus detrainment-temperature feedback. *Geophy Res Lett* 26: 1295–1298

Christy JR (1995) Temperature above the surface layer. *Climatic Change* 31: 455–474

Christy JR (1997) Evidence from the satellite record. In Jones L (ed) *Global warming: the science and the politics* (Vancouver: The Fraser Institute): 55–75

Christy JR, Goodridge JD (1995) Precision global temperatures from satellites and urban warming effects of non-satellite data. *Atmos Environ* 29: 1957–1961

Christy JR, McNider RT (1994) Satellite greenhouse signal. *Nature* 367: 325

Christy JR, Spencer RW, Braswell WD (2000) MSU tropospheric temperatures: Dataset construction and radiosonde comparisons. *J Atmos Oceanic Tech* 17: 1153–1170 (ftp://wind.atmos.uah.edu/msu/t21t/t21tglhmam.do2)

Christy JR, Spencer RW, Lobl ES (1998) Analysis of the merging procedure for the MSU daily temperature time series. *J Clim* 11: 2016–2041

Climate Research Unit (CRU 1999), East Anglia University, United Kingdom (1999); (http://www.cru.uea.ac.uk/cru/data/temperat.htm)

Conrad CP, Hager BH (1997) Spatial variations in the rate of sea level rise caused by the present-day melting of glaciers and ice sheets. *Geophy Res Lett* 24: 1503–1506

Corti S, Molteni F, Palmer TN (1999) Signature of recent climate change in frequencies of natural atmospheric circulation regimes. *Nature* 398: 799–802

Cox PM & al. (1999) The impact of new land surface physics of the GCM simulation of climate and climate sensitivity. *Clim Dyn* 15: 183–203

Cure JD, Acock B (1986) Crop responses to carbon dioxide doubling: a literature survey. *Agric Forest Meteorol* 8: 127–145

Dettinger DM, Ghil M (1998) Seasonal and interannual variations of atmospheric CO_2 and climate. *Tellus* 50B: 1–24

Douglas BC (1992) Global sea level acceleration. *J Geophys Res* 97: 12699–12706

Douglas BC (1995) Global sea level change: determination and interpretation. *Rev Geophys Supplement*: 1425–1432

Drake BG, Leadley PW (1991) Canopy photosynthesis of crops and native plant communities exposed to long-term elevated CO_2. *Plant Cell Environ* 14: 853–860

Fischer H, Wahlen M, Smith J, Mastroianni D, Deck B (1999) Ice core records of atmospheric CO_2 around the last three glacial terminations. *Science* 283: 1712–1714

Forest CE, Allen MR, Stone PH, Sokolov AP (2000) Constraining uncertainties in climate models using climate change detection technique. *Geophs Res Lett* 27: 569–572

Fowler LD, Randall DA (1999) Simulation of upper tropospheric clouds with the Colorado State University general circulation model. *J Geophys Res* 104: 6101–6121

Frei A, Robinson DA (1998) Evaluation of snow extent and its variability in the Atmospheric Model Intercomparison Project. *J Geophys Res* 103: 8859–8871

Gaffen DJ, Rosen RD, Salstein DA, Boyle JS (1997) Evaluation of tropospheric water vapor simulations from Atmospheric Model Intercomparison Project. *J Clim* 10: 1848–1661

Gallo KP, Owen TW, Easterling DR, Jamason PF (1999) Temperature trends of the U.S. Historical Climatology Network based on satellite-designated land use/land cover. *J Clim* 12: 1344–1348

Gifford RM (1992) Interaction of carbon dioxide with growth-limiting environmental factors in vegetative productivity: implications for the global carbon cycle. *Adv Bioclim* 1: 24–58

Glecker PJ & al. (1995) Cloud-radiative effects on implied oceanic energy transports as simulated by atmospheric general circulation models. *Geophys Res Lett* 22: 791–794

Glecker PJ, Weare BC (1997) Uncertainties in global ocean surface heat flux climatologies derived from ship observations. *J Clim* 10: 2764–2781

Godfrey JS & al. (1998) Coupled Ocean-Atmosphere Response Experiment (COARE): An interim report. *J Geophys Res* 103: 14395–14450

Goodridge JD (1996) Comments on regional simulations of greenhouse warming including natural variability. *Bull Amer Meteorol Soc* 77: 3–4

Goodridge, JD (1998) private communication

Goody R, Anderson J, North G (1998) Testing climate models: An approach. *Bull Amer Meteorol Soc* 79: 2541–2549

Gornitz V (1995) Monitoring sea level changes. *Climatic Change* 31: 515–544

Graf H-F, Kirchner I, Perlwitz J (1998) Changing lower stratospheric circulation: The role of ozone and greenhouse gases. *J Geophys Res* 103: 11251–11261

Grove JM, Switsur R (1994) Glacial geological evidence for the medieval warm period. *Climatic Change* 26: 143–169

Grove JM (1996) The century time-scale. In TS Driver, GP Chapman (eds) *Time-scales and Environmental Change* (London: Routledge): 39–87

Hahn CJ, Warren SG, London J (1996) Edited synoptic cloud reports from ships and land stations over the globe, 1982-1991. Tech Rep NDP026B, 45 pp (available from Carbon Dioxide Information Analysis Center, Oak Ridge National Laboratory, Oak Ridge, TN http:// cdiac.esd.ornl.gov/ndp/ndp026b]

Haigh JD (1999) A GCM study of climate change in response to the 11-year solar cycle. Quart J Roy Meteorol Soc 125: 871–892

Hansen J, Lebedeff S (1987) Global trends of measured surface air temperature. *J Geophys Res* 92: 13345–13372

Hansen J, Lebedeff S (1988) Global surface air temperature: Update through 1987. *Geophys Res Lett* 15: 323–326

Hansen J, Ruedy R, Sato M, Reynolds R (1996) Global surface air temperature in 1995; return to pre-Pinatubo level. *Geophys Res Lett* 23: 1665–1668 (http://www.giss.nasa.gov/data/gistemp/GLB.Ts.txt)

Hansen J, Sato M, Ruedy R (1997) Radiative forcing and climate response. *J Geophys Res* 102: 6831–6864

Henderson-Sellers A, McGuffie K, Pitman A (1996) The project for intercomparison of land-surface parameterizaion schemes (PILPS); 1992-1995. *Clim Dyn* 12: 849–859

Henderson-Sellers A & al. (1998) Tropical cyclones and global climate change: a post-IPCC assessment. *Bull Amer Meteorol Soc* 79: 19–38

Houghton JT, Merra Filho LG, Callendar BA, Harris N, Kattenberg A, Maskell K (eds) (1996) *Climatic change 1995: the science of climate change: contribution of working group I to the second assessment.* Report of the intergovernmental panel on climate change. Cambridge: Cambridge University Press

Houghton RA, Davidson EA, Woodwell GM (1998) Missing sinks, feedbacks, and understanding the role of terrestrial ecosystems in the global carbon balance. *Global Biogeochemical Cycles* 12: 25–34

Hu H, Liu WT (1998) The impact of upper tropospheric humidity from Microwave Limb Sounder on the midlatitude greenhouse effect. *Geophys Res Lett* 25: 3151–3154

Hughes MK, Diaz HF (1994) Was there a "Medieval warm period", and if so, where and when? *Climatic Change* 26: 109–142

Idso SB (1989) *Carbon dioxide and global change: earth in transition.* Tempe, AZ: IBR Press

Idso SB (1991) The aerial fertilization effect of CO_2 and its implications for global carbon cycling and maximum greenhouse warming. *Bull Amer Meteorol Soc* 72: 962–965. Idso SB (1991) Reply to comments of L. D. Danny Harvey, B. Bolin, and P. Lehmann. *Bull Amer Meteorol Soc* 72: 1910–1914

Idso SB (1998) CO_2-induced global warming: A skeptic's view of potential climate change. *Clim Res* 10· 69–82

Idso KE, Idso SB (1994) Plant responses to atmospheric CO_2 enrichment in the face of environmental constraints: a review of the past 10 years' research. *Agric Forest Meteorol* 69: 153–203

Idso SB, Kimball BA (1991) Effects of two and a half years of atmospheric CO_2 enrichment on the root density distribution of three year-old sour orange trees. *Agric Forest Meteorol* 55: 345–349

Idso SB, Kimball BA (1994) Effects of atmospheric CO_2 enrichment on biomass accumulation and distribution in Eldarica pine trees. *J Exp Bot* 45: 1669–1692

Idso SB, Kimball BA (1997) Effects of long-term atmospheric CO_2 enrichment on the growth and fruit production of sour orange trees. *Global Change Biology* 3: 89–96

Inamdar AK, Ramanathan V (1998) Tropical and global scale interactions among water vapor, atmospheric greenhouse effect, and surface temperature. *J Geophys Res* 103: 32177–32194

Indermühle A & al. (1999) Holocene carbon-cycle dynamics based on CO_2 trapped in ice at Taylor Dome, Antarctica. *Nature* 398: 121–126

Jain S, Lall U, Mann ME (1999) Seasonality and interannual variations of northern hemisphere temperature: Equator-to-pole and ocean-land contrast. *J Clim* 12: 1086–1100

Jones PD, Osborn TJ, Briffa KR (1997) Estimating sampling errors in large-scale temperature averages. *J Clim* 10: 2548–2568

Karl TR, Knight RW, Christy JR (1994) Global and hemispheric temperature trends: Uncertainties related to inadequate spatial sampling. *J Clim* 7: 1144–1163

Keeling CD, Whorf TP (1997) Atmospheric CO_2 concentrations—Mauna Loa observatory, Hawaii. In trends online: a compendium of data on global change, Carbon Dioxide Information Analysis Center, Oak Ridge National Laboratory (http://cdiac.esd.ornl.gov/ftp/ndp001r7/)

Keeling RF, Manning AC, McEvoy EM, Shertz SR (1998) Methods for measuring changes in atmospheric CO_2 concentration and their application in southern hemisphere air. *J Geophys Res* 103: 3381–3397

Keigwin LD (1996) The little ice age and medieval warm period in the Sargasso Sea. *Science* 274: 1504–1508

Kimball BA (1983) Carbon dioxide and agricultural yield: an assemblage and analysis of 430 prior observations. *Agron J* 75: 779–788

Kimball BA & al. (1995) Productivity and water use of wheat under free-air CO_2 enrichment. *Global Change Biology* 1: 429–442

Knutson TR, Tuleya RE, Kurihara Y (1998) Simulated increase of hurricane intensities in a CO_2-warmed climate. *Science* 279: 1018–1020

Kondratyev KYa (1996) Volcanic eruptions and climate changes. In: Fiocco G, Fua D, and Visconti G (eds) *The Mount Pinatubo eruption's effects on the atmosphere and climate*, NATO ASI series I42 (Berlin-Heidelberg: Springer-Verlag): 273–287

Kuo C, Lindberg CR, Thomson DJ (1990) Coherence established between atmospheric carbon dioxide and global temperature. *Nature* 343: 709–714

Lamb HH (1982) *Climate, history, and the modern world.* New York: Methuen

Landsea CW & al. (1996) Downward trends in the frequency of intense Atlantic hurricanes during the past five decades. *Geophys Res Lett* 23: 1697–1700

Lawlor DW, Mitchell RAC (1991) The effects of increasing CO_2 on crop photosynthesis and productivity: a review of field studies. *Plant Cell and Environ* 14: 807–818

Legates DR, Davis RE (1997) The continuing search for an anthropogenic climate change signal: limitations of correlation-based approaches. *Geophys Res Lett* 24: 2319–2322

Leuliette EW, Wahr JM (1999) Coupled pattern analysis of sea surface temperature and TOPEX/Poseidon sea surface height. *J Phys Oceanography* 29: 599–611

Levis, S, Foley JA, Pollard D (1999) Potential high-latitude vegetation feedbacks on CO_2-induced climate change. *Geophys Res Lett* 26: 741–750

Li Z, Moreau L, Arking A (1997) On solar energy deposition: A perspective from observation and modeling. *Bull Amer Meteorol Soc* 78: 53–70

Liao X, Rind D (1997) Local upper tropospheric/lower stratospheric clear-sky water vapor and tropospheric deep convection. *J Geophys Res* 102: 19543–19557

Lindzen RS (1997) Can increasing carbon dioxide cause climate change? *Proc Natl Acad Sci USA* 94: 8335–8342

Madden RA, Shea DJ, Banstator GW, Tribbia JJ, Weber RO (1993) The effects of imperfect spatial and temporal sampling on estimates of the global mean temperature: Experiments with model data. *J Clim* 6: 1057–1066

Mann ME, Bradley RS, Hughes MK (1999) Northern hemisphere temperatures during the past millennium: Inferences, uncertainties, and limitations. *Geophys Res Lett* 26: 759–762

Marland G, Andres RJ, Boden TA, Johnston C, Brenkert A (1999) Global, regional, and national CO_2 emission estimates from of fossil fuel burning, cement production, and gas flaring: 1751–1996. In trends on line: a compendium of data on global change, Oak Ridge National Laboratory (http://cdiac.esd.ornl.gov/ndp/ndp030.htm)

Mason BJ (1995) Predictions of climate changes caused by man-made emissions of greenhouse gases: a critical assessment. *Contemporary Physics* 36: 299–319

Michaels PJ, Balling Jr RC, Vose, RS, Knappenberger PC (1998) Analysis of trends in the variability of daily and monthly historical temperature measurements. *Clim Res* 10: 15–26

Michaels PJ, Knappenberger PC (1996) Human effect on global climate? *Nature* 384: 522–523

Mortensen LM (1987) Review: CO_2 enrichment in greenhouses. *Sci Hort* 33: 1–25

Murphy JM (1995) Transient response of the Hadley Centre coupled ocean-atmosphere model to increasing carbon dioxide. Part I: control climate and flux adjustment. *J Clim* 8: 36–56

Nerem RS & al. (1997) Improved determination of global mean sea level variations using Topex/Poseidon altimeter data. *Geophys Res Lett* 24: 1331–1334

Palmer TN (1999) A nonlinear dynamical perspective on climate change. *J Clim* 12: 575–591

Parker DE, & al. (1997) A new global gridded radiosonde temperature data base and recent temperature trends. *Geophys Res Lett* 24: 1499–1502

Parker DE, Jones PD, Bevan A, Folland CK (1994) Interdecadal changes of surface temperature since the late 19th century. *J Geophys Res* 99: 14373–14399

Peltier WR (1996) Global sea level rise and glacial isostatic adjustment: an analysis of data from the east coast of North America. *Geophys Res Lett* 23: 717–720

Peng H-F, Wanninkhof R, Feely RA, Takahashi T (1998) Quantification of decadal anthropogenic CO_2 uptake in the ocean based on dissolved inorganic carbon measurements. *Nature* 396: 560–563

Peterson TC & al. (1999) Global rural temperature trends. *Geophys Res Lett* 26: 329–332

Pinter JP & al. (1996) Free-air CO_2 enrichment: responses of cotton and wheat crops. In Koch GW and Mooney HA (eds) *Carbon dioxide and terrestrial ecosystems* (San Diego, CA: Academic Press): 215–249

Polyak I, North G (1997) Evaluation of the GFDL GCM climate variability. 2. Stochastic modeling and latitude-temporal fields. *J Geophys Res* 102: 6799–6812; See also exchanges between North G R (1997) *J Geophys Res* 102: 30161 and Polyak I (1997) *J Geophys Res* 102: 30162

Poorter H (1993) Interspecific variation in the growth response of plants to an elevated ambient CO_2 concentration. *Vegetatio* 104/105: 77–97

Priem HA (1997) CO_2 and climate: a geologist's view. *Space Sciences Reviews* 81: 173–198

Randall D & al. (1998) Status of and outlook for large-scale modeling of atmospheric-ice-ocean interactions in the Arctic. *Bull Amer Meteorol Soc* 79: 197–219

Renno NO, Emanuel KA, Stone PH (1994) Radiative-convective model with an explicit hydrologic cycle. *J Geophys Res* 99: 14429–14441

Robeson SM (1995) Resampling of network-induced variability in estimates of terrestrial air temperature change. *Climatic Change* 29: 213–229

Ross RJ, Gaffen DJ (1998) Comment on Widespread tropical atmospheric drying from 1979 to 1995 by Schroeder and McGuirk. *Geophys Res Lett* 25: 4357–4358

Rossow WB, Cairns B (1995) Monitoring changes of clouds. *Climatic Change* 31: 305–347

Rotstayn LD (1999) Climate sensitivity of the CSIRO GCM: Effect of cloud modeling assumptions. *J Clim* 12: 334–356

Santer BD & al. (1996) A search for human influences on the thermal structure of the atmosphere. *Nature* 382: 39–46; Santer BD & al. (1996) Reply. *Nature* 384: 524

Schimel DS (1995) Terrestrial ecosystems and the carbon cycle. *Global Change Biology* 1: 77–91

Schneider SH (1994) Detecting climate change signals: are there any "fingerprints?" *Science* 263: 341–347

Schroeder SR, McGuirk JP (1998a) Widespread tropical atmospheric drying from 1979 to 1995. *Geophys Res Lett* 25: 1301–1304

Schroeder SR, McGuirk JP (1998b) Reply to Ross and Gaffen 1998. *Geophys Res Lett* 25: 4359–4360

Segalstad TV (1998) Carbon cycle modeling and the residence time of natural and anthropogenic atmospheric CO_2: on the construction of the "greenhouse effect global warming" dogma. In Bate R (ed) *Global warming the continuing debate* (Cambridge: European Science and Environmental Forum): 184–218

Senior CA (1999) Comparison of mechanisms of cloud-climate feedbacks in GCMs. *J Clim* 12: 1480–1489

Siegel DA, Westberry TK, Ohlmann JC (1999) Cloud color and ocean radiant heating. *J Clim* 12: 1101–1116

Simmons AJ, Untch A, Jakob C, Kallberg P, Unden P (1999) Stratospheric water vapor and tropical tropopause temperatures in ECMWF analyses and multiyear simulations. *Quart J Roy Meteorol Soc* 125: 353–386

Soden BJ (1997) Variations in the tropical greenhouse effect during El Nino. *J Clim* 10: 1050–1055

Spencer RW, Christy JR (1990) Precise monitoring of global temperature trends from satellites. *Science* 247: 1558–1562

Spencer RW, Christy JR, Grody NC (1990) Global atmospheric temperature monitoring with satellite microwave measurements: method and results 1979–1984. *J Clim* 3: 1111–1128

Spencer RW, Christy JR (1992) Precision and radiosonde validation of satellite gridpoint temperature anomalies. part I: MSU channel 2 and part II: a tropospheric retrieval and trends during 1979–1990. *J Clim* 5: 847–866

Spencer RW, Braswell WD (1997) How dry is the tropical free troposphere? Implications for global warming theory. *Bull Amer Meteorol Soc* 78: 1097–1106

Stooksbury DE, Idso CD, Hubbard KG (1999) The effects of data gaps on the calculated monthly mean maximum and minimum temperatures in the continental United States: A spatial and temporal study. *J Clim* 12: 1524–1533

Stott PA, Tett SFB (1998) Scale-dependent detection of climate change. *J Clim* 11: 3282–3294

Sun DZ, Held IM (1996) A comparison of modeled and observed relationships between interannual variations of water vapor and temperature. *J Clim* 9: 665–675

Trenberth KE, Christy JR, Hurrell JW (1992) Monitoring global monthly mean surface temperatures. *J Clim* 5: 1405–1423

Valero FPJ & al. (1997a) Atmospheric Radiation Measurements Enhanced Shortwave Experiment (ARESE): Experimental and data details. *J Geophys Res* 102: 29929–29937

Valero FPJ & al. (1997b) Absorption of solar radiation by the cloudy atmosphere: Interpretations of collocated aircraft measurements. *J Geophys Res* 102: 29917–29927

Varejao-Silva MA, Franchito SH, Rao VB (1998) A coupled biosphere-atmosphere climate model suitable for studies of climatic change due to land surface alterations. *J Clim* 11: 1749–1767

Wallace JM, Zhang Y, Renwick JA (1995) Dynamic contribution to hemispheric mean temperature trends. *Science* 270: 780–783

Wallace JM, Zhang Y, Bajuk L (1996) Interpretation of interdecadal trends in northern hemisphere surface air temperature. *J Clim* 9: 249–259

WASA group (1998) Changing waves and storms in the Northeast Atlantic? *Bull Amer Meteorol Soc* 79: 741–760

Weare BC, AMIP group (1996) Evaluation of the vertical structure of zonally averaged cloudiness and its variability in the Atmospheric Model Intercomparison Project. *J Clim* 9: 3419–3431

Weare BC (1999) Combined satellite- and surface-based observations of clouds. *J Clim* 12: 897–913

Weber GO (1996) Human effect on global climate? *Nature* 384: 523–524

Wentz FJ, Schabel M (1998) Effects of orbital decay on satellite-derived lower-tropospheric temperature trends. *Nature* 394: 661–664

Wild M, Ohmura A, Gilgen H, Morcrette JJ (1998) The distribution of solar energy at the Earth's surface as calculated in the ECMWF re-analysis. *Geophys Res Lett* 25: 4373–4376

Wild M, Ohmura A, Gilgen H, Roeckner E (1995) Validation of GCM simulated radiative fluxes using surface observations. *J Clim* 8: 1309–1324

Wunsch C (1999) The interpretation of short climate records, with comments on the North Atlantic and Southern Oscillation. *Bull Amer Meteorol Soc* 79: 245–255

Wylie DP, Menzel WP (1999) Eight years of high cloud statistics using HIRS. *J Clim* 12: 170–184

Yang H, Tung KK (1998) Water vapor, surface temperature, and the greenhouse effect: A statistical analysis of tropical-mean data. *J Clim* 11: 2686–2697

Yao MS, Del Genio AD (1999) Effects of cloud parameterization on the simulation of climate changes in the GISS GCM. *J Clim* 12: 761–779

Yu JY, Mechoso CR (1999) A discussion on the errors in the surface heat fluxes simulated by a coupled GCM. *J Clim* 12: 416–426

Zender CS & al. (1997) Atmospheric absorption during the Atmospheric Radiation Measurement (ARM) Enhanced Shortwave Experiment (ARESE). *J Geophys Res* 102: 29901–29915

Zhang J, Hibler III WD, Steele M, Rothrock DA (1998) Arctic ice-ocean modeling with and without climate restoring. *J Phys Oceanography* 28: 191–217

Zhang MH, Lin WY, Kiehl JT (1998) Bias of atmospheric shortwave absorption in the NCAR Community Climate Models 2 and 3: Comparison with monthly ERBE/GEBA measurements. *J Geophys Res* 103: 8919–8925

Zwiers FW, Kharin VV (1998) Changes in extremes of the climate simulated by CCC GCM2 under CO_2 doubling. *J Clim* 11: 2200–2222